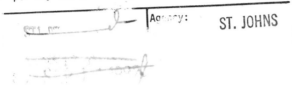

How to Repair Home Laundry Appliances

No. 855
$8.95

How to Repair Home Laundry Appliances

By Ben Gaddis

 **TAB BOOKS**
Blue Ridge Summit, Pa. 17214

Preface

There's nothing really frightening about repairing a washer or dryer. These appliances contain very ordinary mechanical and electrical components that perform simple repetitive tasks in response to dialed or timed commands. Yet, repairable motors are replaced every day by servicemen who'd rather sell a costly new one than spend the time it takes to fix the old. It is the purpose of this book to give you all the know-how necessary to make you feel at ease and self-confident when you're faced with any washer/dryer problem, whether it's an installation requiring a wiring modification or an appliance that's refused to operate.

If you've a fair background in electricity or electronics, this book will prove the only tutor you need to actually set up shop—to go into the home laundry service business on a full- or part-time basis. If you're a stranger to the electron, you'll find enough information in this book to help you through any washer or dryer problem you're ever apt to face. You have to learn a few simple rules—what you can and can't do, how to troubleshoot using progressive and logical procedures, how to spot basic appliance problems using your own natural senses—but before long you'll understand the operational differences between your machines' timed functions, and you'll have a "handle" on potential problems even before they develop enough momentum to put your laundry out of commission.

If you're already doing washer and dryer repair for a living, you'll find this book helpful in expanding your present knowledge of laundry appliances. If you're a self-styled handyman and like to fend for yourself, you'll appreciate the simplified explanations of circuit functions and the detailed schematics that accompany them.

Ben Gaddis

Contents

Chapter 1

Introduction

The modern home laundry is more than just a convenience. Any housewife will tell you that her washer and dryer are her most prized appliances. This should be no surprise. Modern washers and dryers with cycles for any fabric, along with today's knit and permanent-press fabrics, eleminate much of the washday drudgery of ten years ago. And ironing, once a time-consuming and difficult chore, has been greatly simplified. In view of this it is easy to understand why most housewives are upset when they have to call the appliance serviceman. What could be more frustrating than a waist-high stack of dirty laundry on Monday morning with the washer gone mad and the laundry room ankle deep in sudsy water? Or when, on the coldest, wettest day of the year and with two tubs of wet clothing, the dryer tumbles just fine but doesn't get hot enough to melt an ice cube?

Then comes the worry about how long repairs will take and how much they will cost. At today's prices these worries are understandable. Increases in the cost of parts, labor, and general operation have caused the cost of appliance repair to climb along with prices of other goods and services. In some areas of the country parts are often difficult to obtain. This can have a great bearing on the time required to make repairs.

Considering experiences such as these, the irritation felt by the owner of a defective appliance seems only natural.

Keep this in mind during your contacts with customers. How you handle the first few moments of the service call can have a great effect on the customer's attitude towards you and your services. Don't make promises that you can't keep, but be as reassuring and as helpful as you can. This will relieve much of the tension sometimes experienced between the customer and the serviceman.

Let's start with the first contact—usually a phone call. Just how you answer the call can have considerable effect on the customer. Assume a friendly, cheerful voice to determine the nature of the complaint. Find out as much as you can about the equipment: make, model, and the approximate date of purchase. This information will help you determine what parts and special tools should be taken on the service call. Establish with the customer a definite time you will arrive. Don't add to the aggravation by keeping the customer waiting all morning when you could possibly establish a time to within an hour.

A well-equipped service truck represents a large investment, and the selection of parts to have available in the truck is sometimes difficult to determine. This is extremely important, however. Not only will a well-stocked service truck save you time and provide a better profit margin, but it means that appliances will not be out of service for too long. Experience shows that most customers don't mind the repair bill as much as they mind the appliance being out of service while you wait for parts that must be ordered. The selection of parts you carry in the service truck and stock at the shop will depend on several factors. Perhaps the most obvious factor is the make of appliances you normally service. The decision must ultimately be made from your own experience.

Your conduct, attitude, and appearance will make an immediate impression on your customer. If you are polite and helpful with a professional attitude and professional equipment, you are off to a good start. A sloppy appearance or a casual attitude can destroy the customer's confidence in your ability to service complex equipment.

A profitable home laundry service must be more than just a fix-it business. These machines represent a large investment by the consumer, and he rightly expects high-quality service for the price he must pay. This book can help you render that professional service.

HOME LAUNDRY CENTER

Modern home laundries include automatic washers and dryers (Fig. 1-1) and are a very important part of today's home. Most homes today are built with a specific area set aside for the home laundry. Adequate water, drainage, electrical power, and a gas supply are provided for these areas.

The washer in Fig. 1-1 is completely automatic and includes special cycles for delicate and permanent-press fabrics. A presoak cycle and a lint filter are also featured. A water hookup behind the washer furnishes hot and cold water. A temperature control blends hot and cold water to achieve the desired temperature. A level control fills the washer with the correct amount of water for the load to be washed. A proper electrical service of at least 15A is available behind the washer. Although not part of the washer itself, a level floor provides proper support for convenience and safety.

Fig. 1-1. A modern home laundry includes an automatic washer and dryer in a convenient location.

A 240V outlet behind the electric dryer supplies power for heat and automatic operation. A flexible vent pipe connects the dryer exhaust to a vent installed through the wall. This expels heat and lint fibers from the machine to the outside atmosphere. In addition to the requirements of the machines themselves most installations include storage space for soiled clothing, laundry supplies, and ironing equipment.

Many laundries in older homes do not feature all the conveniences discussed here. Those that do may have conveniences that are improperly installed. Poor plumbing and venting and dangerously inadequate wiring are sometimes encountered. For this reason we will discuss the wiring needs of modern home laundries. This is not to teach you to wire the laundry but to recognize proper and improper wiring. The appliance serviceman is frequently the first professional familiar with electrical equipment with whom the homeowner discusses his laundry wiring problems.

One word of caution: Most states, counties, and cities prohibit any additions, alterations, or modifications to home wiring by anyone other than a licensed electrician. You are cautioned against offering such service without first consulting your local government offices. The city (or county) engineer can give you the rules for your locality.

ELECTRICITY

Although some of the electricity we use is generated locally, most of it comes to us through a very complex distribution system. Even electricity generated for special purposes must be routed through some sort of distribution system. Let's take a look at the system that brings electricity into the home.

Generation

Electricity in the United States has been generally standardized—the electricity used in Boston has the same characteristics as that used in Omaha or Seattle. Two values of voltage are usually supplied to modern residences—120V and 240V. The voltage may vary slightly about these values, depending on the particular distribution system, but never more than a few volts. Frequency is set at 60 hertz (Hz)—60 complete cycles each second. This voltage may be generated in several ways. Some cities have their own generating plants while others, especially smaller towns, bring power in from some remote location. The generator may be hydroelectric or steam turbine, depending on the resources available.

Large hydroelectric dams are located in many areas of the United States where there is an abundance of rivers and streams. The Northwestern United States and the Central States east of the Mississippi River are prime examples. Generators in these dams convert mechanical motion to electricity. The dam itself blocks the natural flow of the stream and causes the water to back up, forming a large man-made lake. Near the dam this water is quite deep and exerts and extremely high pressure on anything near the bottom. That pressure, with the natural downward flow of the water, turns the generators.

Steam turbine generators are similar in some respects to hydroelectric generators. The main differences are in the turbine itself and its drive mechanism. Boilers are used to generate the steam that spins the generator. Fuel for the boiler will vary with geographic area.

Conductors

The voltage is brought to the home from the generators by way of conductors. Conductors vary in size, shape, insulation, and other qualities, depending on their intended use. Copper and aluminum have become the most common metals used as conductors. Aluminum has the advantages of weighing and costing less than copper. It also has the disadvantage of lower current-carrying capabilities, requiring a larger conductor to carry the same amount of current.

Regardless of the type of material used in making the wire, all conductors have certain common properties with which you should be familiar. All material has a specific resistance, measured in ohms and symbolized by the Greek letter omega (Ω). Resistance is different for every material and is usually quoted as the amount of resistance present in a circular-mil foot of the material. An example of the

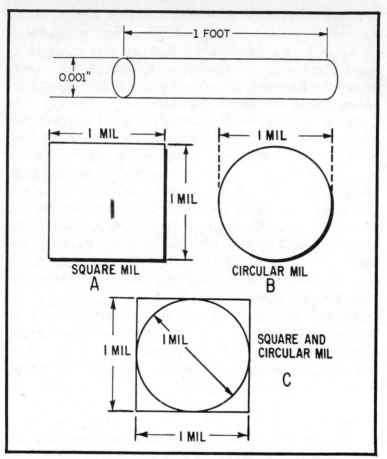

Fig. 1-2. The circular-mil foot.

circular-mil foot is shown in Fig. 1-2, the conductor being one mil in diameter and one foot in length.

By using the specific resistance per foot, the resistance of any length or diameter of a material can be computed as

$$R = \frac{\rho L}{A}$$

where

R = resistance in ohms

ρ = specific resistance in ohms per circular-mil foot of the material

L = length of the conductor in feet
A = cross sectional area in circular mils

From this formula you can see that as the length of a conductor is increased, the resistance is increased; and if the cross sectional area of the conductor is increased, the resistance is decreased. Resistance then varies inversely with the cross sectional area and inversely with the square of the diameter (since area $= D^2$). Therefore, if the diameter is halved, the area is quartered and resistance is quadrupled. If the diameter is doubled, the area is quadrupled and resistance is quartered.

The formula shows that a small wire will offer more resistance to current than a large wire of the same material. Since power is computed as $P = I^2 R$ this increase in resistance causes the smaller wire to consume more power than a larger wire carrying the same current. This power is dissipated as heat. Should the current become high enough the heat will become so intense that the insulation and the wire will melt. This is why large conductors must be used to carry high currents.

The loss of power in a conductor also means a loss of voltage (E) at the load. Assume a circuit (Fig. 1-3A) that has a 110V generator supplying power to a 100Ω load through two wires having 5Ω resistance each. The generator sees a total resistance (R) of 110Ω, the resistance of each wire *plus* the load resistance. Under these conditions a total current (I) of 1A is present in the circuit: $I = E/R = 110V/110Ω = 1A$.

With 1A of current the three resistances will cause voltage drops around the circuit as shown in Fig. 1-3B. As can be seen, the voltage across the load will be 100V instead of 110V. Although this particular example is somewhat exaggerated, it illustrates the results of using a conductor that is too small in diameter or too long in length. This condition causes the load (appliance in our case) to operate improperly—underheating elements, overheating or burning motors, etc. From this example we can see the importance of using the proper wire for the job.

Wire size has been standardized throughout the United States. The size is given as a number; the smaller the number

19

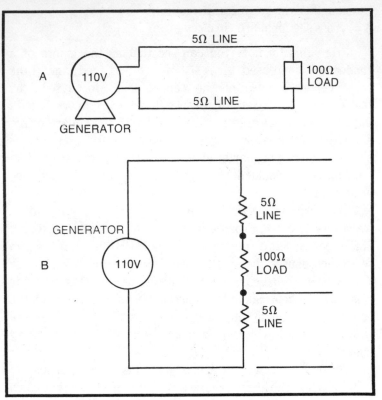

Fig. 1-3. Voltage loss in an electrical conductor

the larger the size. Common sizes run from 0000, which has a cross-sectional area of 0.166 sq in., down to 40 gage, having an area of 0.0000078 in. A conductor may be one solid wire or a group of smaller wires twisted together. The twisted strands are used where a large conductor with good flexibility is needed.

In addition to wire size, the insulated covering is also important. The conductor may be a single insulated wire or many conductors in one cable, each separated from the other by its own insulated covering. In some cases the cable is covered with metal braid (armor) to protect the insulation. Other installations require the conductors to be run through a metal pipe called conduit.

Most electrical distribution systems make use of a wiring code in which the insulated covering of each conductor is colored according to its use. This is especially true in residential and industrial applications.

Neighborhood Distribution

Electrical power is brought into the home from the generating source by high-tension wires. At the generating station the voltage output is typically 10,000–20,000V. In order to avoid current losses the voltage is increased through step-up transformers to somewhere between 20,000 and 150,000V. By transmitting the power at a very high voltage and minimum current power, losses in the wire are reduced. This allows the use of smaller wire.

From the step-up transformers the electrical power is carried to local substations where the voltage is reduced and routed to the surrounding neighborhoods. In some areas this is done by the familiar utility poles while in other areas an underground transmission system is used. Located throughout the substation area are transformers that step down the voltage to a level where it can be used in the home. You have probably seen these transformers at the tops of utility poles in some areas.

Figure 1-4 illustrates the function of the utility pole transformer. The power from the substation is routed to the local step-down transformers, where the high voltage is

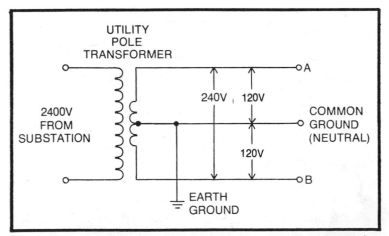

Fig. 1-4. The neighborhood electrical distribution system uses an underground transmission system or utility poles.

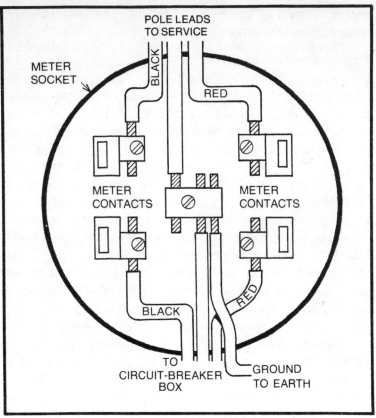

Fig. 1-5. The meter hookup includes contacts that plug into a socket.

reduced to two values—120V and 240V. Some older homes have only two-wire service, the neutral and one leg of the secondary. Newer homes have three-wire service, the neutral and both legs of the secondary. The voltage between the neutral and any one leg will be about 120V, while 240V is present between the two "hot" legs. Since the National Electric Code (NEC) requirements have brought changes in service entrance requirements we concentrate our efforts on the 240V installation.

Service Entrance

The power is brought from the pole to the service entrance meter. The meter is equipped with contacts that plug into a socket similar to that shown in Fig. 1-5. Three conductors are used in a typical 240V installation. Between the red and black leads coming from the pole will be approximately 240V. The voltage between either the black or red and the neutral will be about 120V. The contacts on the meter connect the pole leads to leads entering the circuit-breaker box. All current used is monitored by the meter. Notice that the neutral is grounded to earth by a grounding conductor.

Heavy-gage conductors are used to connect the meter socket to the circuit-breaker box. This box acts as the distribution point from which all circuits are routed. In some older installations, fuses will be used instead of the more convenient circuit breakers. The fuse must be replaced if a circuit is overloaded. If an overload occurs in a circuit protected by a circuit breaker, the breaker need only be reset once the overload condition has been corrected.

If you traced the electrical circuit in a modern home from the point where it enters the home, immediately after the meter you would come to the distribution box or breaker box. This is the point from where each circuit is routed to its area of use. The NEC has specific rules which must be followed when these circuits are installed. Although you will not likely make any changes or additions to these circuits (in most cases the law forbids changes or additions) you should, for safety, be familiar with code requirements. The heavy loads placed on the electrical system by home laundry equipment make it necessary for the appliance technician to know enough about home wiring systems to recognize inadequate or unsafe wiring. The breaker box is one of the most important parts of the home wiring system. Before examining the breaker box itself let's look briefly at the general requirements placed on the home wiring system by the NEC.

The lighting circuit must provide a minimum of three watts of lighting for each square foot of living space. That's just for lighting, and the bare minimum. Appliance requirements are in addition to the lighting load. Two individual 20A, 1500W appliance circuits must be provided. These circuits are separate from, and in addition to, any circuit for a special appliance, such as an electric range or water heater. You may not find all these features in many

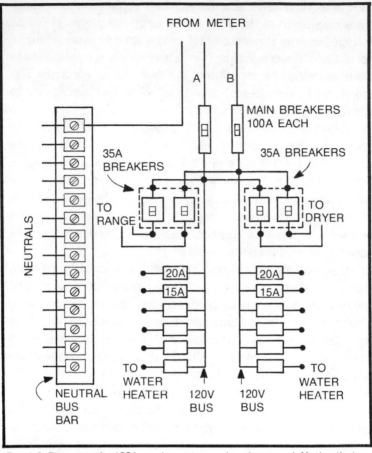

Fig. 1-6. Diagram of a 100A service-entrance breaker panel. Notice that no overload devices are found in the neutral lines.

older homes, but any new construction must comply with these requirements.

Service entrances, and therefore breaker boxes, are identified by the maximum amount of current to be supplied through that entrance. In the early days of home electrical service, installations were as small as 30A. The 60A and 100A services are more typical today. Some entrances are capable of handling 160A. The diagram in Fig. 1-6 depicts a 100A service-entrance breaker panel. The neutral lead is brought in and tied to the neutral tie for all circuits. Notice that no fuse or breaker is found in the neutral leg of any circuit. This is required by the NEC. The neutral must be unbroken throughout the entire wiring system.

The hot wires (red and black) run from the meter socket to the main circuit breaker or fuse. Although separate 100A fuses or circuit breakers are provided for each leg, they are mechanically connected (ganged) so that if one circuit is opened by tripping the breaker (or breaking the fuse) the other circuit will also be interrupted. Each of these 100A breakers supplies its own bus, from which the various branch circuits emerge. Notice that between any hot leg and neutral there is 120V, while between the two hot legs there is 240V. For any one of the lighting circuits it will be necessary to run one neutral wire and one hot wire. The neutral ties directly to the neutral bus while the hot wire is connected to one of the hot buses through a circuit breaker.

In addition to its separate 120V bus, each 100A breaker supplies one leg of the 240V circuits for the electric range and electric dryer. Both these appliances are protected by ganged 35A breakers in each 120V leg. Similar to the main breaker, these special breakers open both legs to the appliance simultaneously. Again, an unbroken neutral is supplied from the neutral bus. The water heater receives its power through two separate 15A breakers.

Each of the lighting and convenience circuits are protected by a 15A circuit breaker. These circuits supply electrical power for room lighting and such items as TVs, radios, and other devices. Each of these circuits should be limited to no more than ten outlets, but in no case should the load be more than 15A.

The NEC requires that all new installations include two separate 20A appliance circuits. These circuits are not to be used for any purpose other than appliance operation. In our example in Fig. 1-6, the top breakers on each bus provide these circuits.

One important aspect of using the distribution box is balancing the load. Circuits should be connected across the supply buses so that they balance or equalize the load on each of the 120V hot buses as much as possible. The importance of this is illustrated in Fig. 1-7.

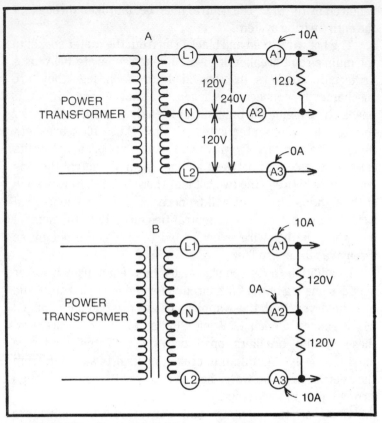

Fig. 1-7. Balancing the load of the secondary of a power transformer.

Assume that the windings in Fig. 1-7A represent the secondary of a power transformer whose output is 240V between L1 and L2, 120V between N and L1, and 120V between N and L2. An ammeter has been inserted in each leg for illustration. Now, if a load having 12Ω is placed across the 120V circuit between N and L1 the current will be 10A: $I = E/R = 120V/12\Omega = 10A$.

Under these conditions with no load, and therefore no current flow across N to L2, ammeter A1 will read 10A; A2, 10A; and A3, 0A. This means that if we ignore the power losses in the wire itself 10A flows through both the L1 and N conductors.

Now let's add a second 12Ω load, but add this new resistance between N and L2 as shown in Fig. 1-7B. This load would also draw 10A as indicated by A3. The effect of the current in the N leg is what we are interested in here. Notice that with the two loads connected as shown we have a 24Ω circuit connected across 240V between L1 and L2. Since these loads are equal each must drop half the total voltage. This means that half the voltage is dropped across the load between N and L1, and the other half is dropped across the load between N and L2. Theoretically, we see that the voltage at the junction of the two loads (the N conductor) will be zero and there would be no current flow through N. This assumption illustrates how a balanced load can reduce current flow in the neutral conductor. Since any conductor has some resistance, current through that resistance causes voltage drops along the line. If current in the neutral conductor is reduced by balancing the load, line losses can be kept to a minimum.

Most contractors are careful to balance loads in new electrical installations. You may encounter some systems where homeowners have done their own add-on work to the original circuits, however. These may or may not be according to the NEC and they may or may not be balanced. You should be extra careful when working with such circuits since you have no assurances that proper safety precautions have been taken.

In many cases older homes have been remodeled and, to save on remodeling costs, the existing wiring was retained. The extra demand placed on the existing service, however, requires the addition of new circuits. In some cases the breaker box has been replaced while in others the extra power is provided from a small add-on unit. Figure 1-8 illustrates how such add-ons are installed. The installation shown was originally a 60A service. Note the similarities between the 60A breaker box in Fig. 1-8 and the 100A box shown in Fig. 1-6.

A heavy conductor is connected between the neutral bus in the existing box and the new box. In this particular case the 35A range circuit was unused. Therefore, the additional breaker box can be added by connecting it to the range breaker. Two 15A circuits can be added to each hot leg of the 240V range circuit.

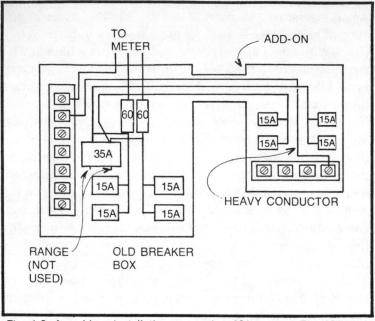

Fig. 1-8. An add-on installation converting 60A service. The 60A main breakers must be changed to 100A.

General Home Wiring

The importance of proper wiring was mentioned earlier. Wiring too small in diameter or too long causes excessive line losses. At best, incorrect wiring can cause inefficient operation of appliances; at worst, shocks and fire hazards. From the efficiency standpoint alone you should be concerned about wiring that supplies power to appliances you must service. There are two different home wiring systems you may encounter in your service work.

You will find the two-wire system in older homes and in some illegal do-it-yourself modifications. Installation of this system is no longer allowed under NEC rules. Figure 1-9 shows how a two-wire circuit is run from a circuit-breaker box to a bedroom.

Notice that one wire is brought from the neutral bus to a junction box where the lighting and convenience outlets in this branch are tied together. By code requirements the insulating wrapper of this neutral conductor must be white. From the connection box the white neutral wire is supplied directly to the lighting fixture, and one white wire ties all the convenience outlets together in parallel.

The hot wire is run from the circuit breaker to the junction box and parallels the convenience outlets. Notice that the lighting circuit is different, however. Instead of going directly to the lighting fixture, as did the neutral, the hot wire goes first through a switch then to the lighting fixture. The NEC requires that the hot wire be covered with a black insulative wrapper. These black and white color codings must be maintained throughout the system.

Although a switch in the neutral line could be made to control the lighting fixture, the code requires that no switches, breakers, or any other device capable of interrupting the neutral circuit be installed. Notice in Fig. 1-9 that without any

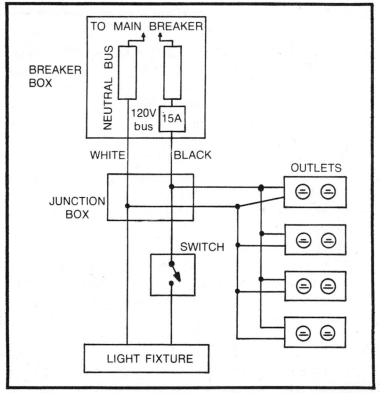

Fig. 1-9. Typical two-wire branch circuit running from the breaker box to the bedroom. The NEC requires that all "hot" wires be covered with black insulation. The neutral wire is covered with white insulation.

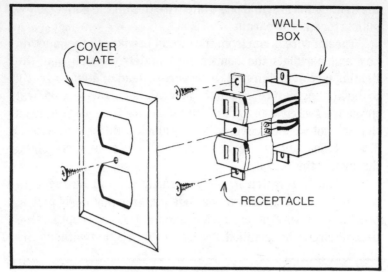

Fig. 1-10. The two-wire convenience outlet.

breakers or switches the neutral terminal on the lighting fixture is common (zero resistance) with the neutral terminals of the convenience outlets. Also. all the neutral terminals of the wiring system are common since all neutrals are uninterrupted and tie to the neutral terminal block or bus in the circuit-breaker panel.

If you traced the white neutral back to the service entrance you would find that this wiring is grounded through the service-entrance ground. Here a long metal stake or plate is sunk deep into the earth, and the neutral leg is tied to that grounding wire.

The quickest and most obvious way to recognize the two-wire circuit is through examination of the convenience outlets. Figure 1-10 illustrates the common two-wire convenience outlet. One black and one white wire bring the neutral and hot legs to opposite sides of the outlet. There may be more than two wires in the box, however. These wires would parallel the remaining outlets on that branch. Notice that the metal box housing the outlet is not connected in any way to the electrical system.

A variation of the two-wire system can be identified by the type of two-wire outlet used. More recent two-wire installations use a polarized outlet. These outlets are similar to the two-wire units just discussed except for the physical size of one of the connectors in the outlet. Close examination of the polarized outlet will reveal that one of the connectors is slightly larger than the other. When polarized outlets of this type are used the neutral wire should be connected to the side of the outlet that is common to the larger connector.

The polarized outlet came into being with appliances that have polarized power cords. These cords have one prong on the plug larger than the other. Most of their applications have to do with electronic equipment, such as radios and TVs, and have little to do with our subject. But you should be able to recognize these installations when you see them. Keep the following in mind:

- Under no circumstance should polarized connectors or outlets be modified to defeat the polarization feature. This can result in severe shock hazards to those who operate appliances connected to these circuits.
- Never assume that polarized outlets are wired properly. Test the circuit to insure that the neutral leg is properly wired.
- Never trust wiring color codes. In spite of NEC rules, you may encounter systems where safety precautions have been violated or ignored.

The modern method of home wiring, and the only one permitted by the present code in new installations, is called the three-wire system. This system can be identified by the features shown in Fig. 1-11.

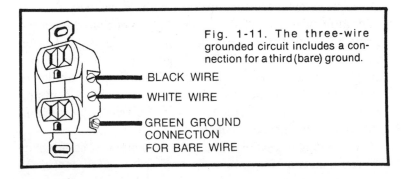

Fig. 1-11. The three-wire grounded circuit includes a connection for a third (bare) ground.

BLACK WIRE

WHITE WIRE

GREEN GROUND CONNECTION FOR BARE WIRE

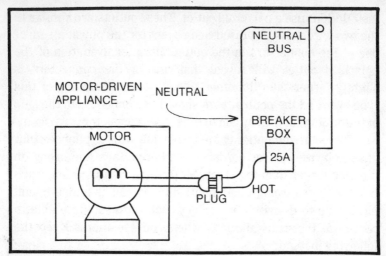

Fig. 1-12. The dangers of ungrounded appliance circuits include shock and fire hazards.

Wiring for this system uses the black and white color-code system, as used in the two-wire system, plus an additional wire. The third wire is not insulated and is called the *bare ground*. Outlets for these systems have three connections: the hot wire, the neutral, and a ground. This bare wire is tied to the ground connector and the metal boxes throughout the system. Code requirements state that this ground, similar to the neutral, must be unbroken throughout the entire system.

The ground wire is sometimes referred to as the *nonconducting* neutral since under normal operating conditions no current flows in this wire. Notice that nowhere in the system is the ground connected in any way to either the neutral or hot sides of the circuits. Instead, somewhere near the service entrance the ground wire is connected to a cold-water pipe or a grounding rod that is separated from the grounded neutral. Hot water, gas, or plastic pipes cannot be used for the grounding device. In some areas where plastic pipe is used as the cold-water line, a grounding rod similar to that employed with the neutral must be used. This is common in rural areas where plastic pipe is used between the well and the cold-water entrance to the house.

Grounding the system offers the advantage of safety over the common two-wire system. One important safety feature is illustrated in Fig. 1-12. A motor-driven appliance with a metal cabinet is shown connected to a two-wire 120V system by a two-wire power cord and a two-prong outlet. Notice how the neutral wire and the hot wire connect to the breaker panel. As long as the insulation and electrical connections in the appliance remain in good shape there is no problem.

But what happens if one of the hot wires becomes disconnected or the insulation breaks down and the hot side of the line becomes tied to the metal cabinet? A dangerous situation evolves but no permanent damage is sustained.

Everything remains normal and no external symptoms of trouble appear until some housewife attempts to use this appliance. If she is standing on an uninsulated floor or is accidently touching another metal object, there is an electrical path from the 120V, 20A breaker through her body to the object she is touching. Severe electrical shock or even death could be the result. Had this appliance been properly connected to a three-wire circuit, the possibility of this accident ever occurring may not have been completely eliminated, but the degree of danger would have been lowered.

Look at Fig. 1-13. Same appliance, same defective hot lead, only this time the cabinet is connected to a three-prong grounded outlet. The ground connector is then tied to the system ground by the bare conductor explained with Fig. 1-10.

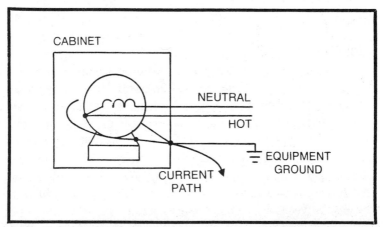

Fig. 1-13. A three-wire grounded appliance provides better safety against potential hazards.

Under these conditions current can flow from the 20A breaker through the faulty connection to the cabinet of the appliance. Since the cabinet is connected to ground through the three-wire system, current will start to flow before anyone or anything comes in contact with the cabinet. Should this current become heavy enough, such as cases involving direct shorts, the circuit breaker will trip and no power will be present at the outlet. Although this doesn't entirely remove the possibility of electrical shock from a defective appliance, it does decrease the danger significantly.

The need for adequate grounding of the home electrical system cannot be overemphasized. The three-wire scheme provides for protection provided the wiring rules of the NEC are followed. As an added safety, however, it is good practice to provide an additional ground for each major stationary appliance. The ground applied to the third wire of the system is called the system ground to distinguish it from the equipment (individual) ground. The code requires that the resistance between the system ground and equipment ground be less than 25Ω.

Individual equipment grounds can be attached from metal parts of the cabinet to a good ground such as a cold-water pipe. This added protection should be provided when such appliances are installed in damp spaces, such as basements, or when there is any doubt as to the quality of the system ground.

Wire Types and Sizes

Types and sizes of wire used in home wiring should be recognized by any technician working with major electrical appliances. You should know enough in this area to advise your customers concerning the wiring of their homes, especially the appliance circuits.

Three types of wire are used in home wiring and are illustrated in Fig. 1-14. *Conduit* wiring (Fig. 1-14A) consists of thin pipe through which the wire is run. The conduit protects the insulative wrapping of the conductors against moisture and abrasion. There are specific limits as to the size and number of conductors that may be carried in any particular size of conduit. These details are not required for our purposes here, however. Properly installed conduit wiring does not require a separate ground since the metal piping should be well grounded.

Similar to, and sometimes confused with, conduit is *metallic shielded cable* (Fig. 1-14B). This offers many of the advantages of conduit and is somewhat easier to run. It is available in several sizes coded by number and size of conductors. The metallic shielding of this cable can be used for a grounding conductor.

One of the most popular types of wiring in use today is the *nonmetallic sheathed cable* shown in Fig. 1-14C. Nonmetallic cable is more flexible and much lighter than sheathed cable or conduit. Depending on where it is to be installed this cable may be weatherproofed for outdoor, underground, or indoor use. There are also types for use in hot, humid, or extremely dry areas.

This cable also comes in different sizes coded by the number of conductors and the wire size. A group of letters is used to designate the type of cable according to the area of intended use. For example, TW is thermoplastic cable for general use in wet areas. Size is designated by a combination of numbers. The number 10-2, for example, is used to designate a cable containing two conductors of 10-gage wire. Three 12-gage conductors contained in a single sheath would be

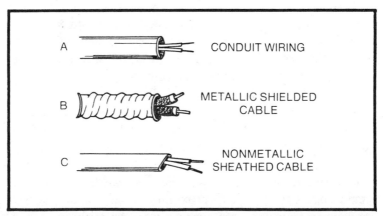

Fig. 1-14. The general types of electrical cables include conduit, metallic shielded, and nonmetallic sheathed cables.

designated 12-3. Nonmetallic cable comes with or without the separate ground conductor. The ground conductor does not have an insulative wrapper.

Wire size is as important as the insulative qualities of the wrapper. Common American wire sizes are designated by numbers—the larger the number, the smaller the diameter of the wire. The smallest wire size allowed in home wiring is 14-gage. Figure 1-15 is a table of common wire sizes as determined by the American Wire Gage (AWG) standard accepted by the NEC. Figure 1-16 is a typical gage used to measure the diameter of wires.

The length and cross-sectional area must also be considered when selecting the proper wire size. The cross-sectional area (gage) must be increased if the length of the run is increased and the current is held constant. This is because of the increased voltage drop across the line.

Appliance Circuits

The requirements placed on lighting and convenience circuits apply to appliance circuits as well. In addition, there are several more requirements. First let's consider the 240V appliance circuits, in particular the water heater and electric dryer circuits. You may recall from the discussion of the service entrance that both the electric range and electric dryer have their own circuits, complete with a fuse or circuit breaker. The water heater has its own separate branch taken from the 120V buses. Under no conditions should any other circuit be tied into these circuits, not only for the obvious reason that they are 240V circuits but also because the water heater circuit is not supplied with a neutral wire.

Automatic washers and gas dryers require 120V for operation. These should be at least 20A branch circuits. In addition to having three-prong grounded outlets it is good practice that these appliances be equipped with an equipment ground. No effort should be spared to make the laundry equipment as safe as possible. The inherent danger of water and electricity in such proximity requires the service technician to insure safe operation by the customer. When servicing laundry appliances always check the grounding system and general condition of the wiring as a part of your service call.

ELECTRICAL SAFETY

Many electrical appliance operators are not aware of the safety precautions that should accompany their operation. You do your customers a great disservice when you do not instruct them in the safe operation of their appliances. When we study the installation of these appliances, we will see the details of safe home laundries. For now, let's look at some of the safety devices in the appliance branch circuits. Proper grounding has already been discussed and will be brought to mind again in later chapters.

Fuses

You will encounter different types of fuses (see Fig. 1-17) in home wiring systems. The most common is a plug fuse. These fuses range in link size up to 30A. The threaded base of these fuses is the same size as an ordinary incandescent lamp. As seen in the illustration, there are three basic parts to the fuse: the shell screw, the centerpost, and the fusible link. The fusible link connects the centerpost to the shell screw. When a load draws current higher than the rating of the fuse the heat generated by the current will melt the fusible link.

These fuses have several disadvantages, not the least of which is the danger of overfusing. Frequently, when a customer experiences repetitive blown fuses on a particular branch circuit, he will overfuse (use a larger capacity fuse) in the fuse holder. Occasionally you may find the fuse defeated by a conductor, such as a penny, inserted behind the fuse. Overfusing is a very dangerous practice and should be pointed out to the customer whenever encountered.

There are times when it is desirable, however, to allow brief high currents in certain branch circuits. For example, circuits supplying electric power to motors are subject to momentary high currents needed to start the motor. These currents would blow the ordinary fuse. To prevent this, time-lag or slow-blow fuses (shown in Fig. 1-18) are used. Instead of melting the fusible link, a constant overload causes

Gage number	Diameter (mils)	Cross section		Ohms per 1000 ft.		Ohms per mile 25° C. (= 77° F.)	Pounds per 1000 ft.
		Circular mils	Square inches	25° C. (= 77° F.)	65° C. (= 149° F.)		
0000	460.0	212,000.0	0.166	0.0500	0.0577	0.264	641.0
000	410.0	168,000.0	.132	.0630	.0727	.333	508.0
00	365.0	133,000.0	.105	.0795	.0917	.420	403.0
0	325.0	106,000.0	.0829	.100	.116	.528	319.0
1	289.0	83,700.0	.0657	.126	.146	.665	253.0
2	258.0	66,400.0	.0521	.159	.184	.839	201.0
3	229.0	52,600.0	.0413	.201	.232	1.061	159.0
4	204.0	41,700.0	.0328	.253	.292	1.335	126.0
5	182.0	33,100.0	.0260	.319	.369	1.685	100.0
6	162.0	26,300.0	.0206	.403	.465	2.13	79.5
7	144.0	20,800.0	.0164	.508	.586	2.68	63.0
8	128.0	16,500.0	.0130	.641	.739	3.38	50.0
9	114.0	13,100.0	.0103	.808	.932	4.27	39.6
10	102.0	10,400.0	.00815	1.02	1.18	5.38	31.4
11	91.0	8,230.0	.00647	1.28	1.48	6.75	24.9
12	81.0	6,530.0	.00513	1.62	1.87	8.55	19.8
13	72.0	5,180.0	.00407	2.04	2.36	10.77	15.7
14	64.0	4,110.0	.00323	2.58	2.97	13.62	12.4
15	57.0	3,260.0	.00256	3.25	3.75	17.16	9.86

Gage No.							
16	51.0	2,580.0	.00203	4.09	4.73	21.6	7.82
17	45.0	2,050.0	.00161	5.16	5.96	27.2	6.20
18	40.0	1,620.0	.00128	6.51	7.51	34.4	4.92
19	36.0	1,290.0	.00101	8.21	9.48	43.3	3.90
20	32.0	1,020.0	.000802	10.4	11.9	54.9	3.09
21	28.5	810.0	.000636	13.1	15.1	69.1	2.45
22	25.3	642.0	.000505	16.5	19.0	87.1	1.94
23	22.6	509.0	.000400	20.8	24.0	109.8	1.54
24	20.1	404.0	.000317	26.2	30.2	138.3	1.22
25	17.9	320.0	.000252	33.0	38.1	174.1	0.970
26	15.9	254.0	.000200	41.6	48.0	220.0	0.769
27	14.2	202.0	.000158	52.5	60.6	277.0	0.610
28	12.6	160.0	.000126	66.2	76.4	350.0	0.484
29	11.3	127.0	.0000995	83.4	96.3	440.0	0.384
30	10.0	101.0	.0000789	105.0	121.0	554.0	0.304
31	8.9	79.7	.0000626	133.0	153.0	702.0	0.241
32	8.0	63.2	.0000496	167.0	193.0	882.0	0.191
33	7.1	50.1	.0000394	211.0	243.0	1,114.0	0.152
34	6.3	39.8	.0000312	266.0	307.0	1,404.0	0.120
35	5.6	31.5	.0000248	335.0	387.0	1,769.0	0.0954
36	5.0	25.0	.0000196	423.0	488.0	2,230.0	0.0757
37	4.5	19.8	.0000156	533.0	616.0	2,810.0	0.0600
38	4.0	15.7	.0000123	673.0	776.0	3,550.0	0.0476
39	3.5	12.5	.0000098	848.0	979.0	4,480.0	0.0377
40	3.1	9.9	.0000078	1,070.0	1,230.0	5,650.0	0.0299

Fig. 1-15. Standard wire sizes are measured by the American Wire Gage code.

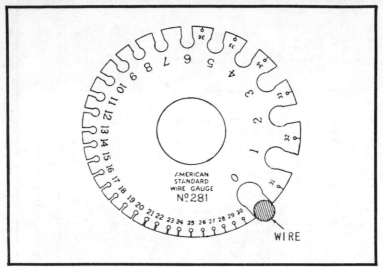

Fig. 1-16. A typical gage used to determine the diameter of a wire.

the solder to melt and release the spring which opens the circuit. Motor currents do not last long enough to melt the solder. Therefore, these fuses can handle short periods of high current required in motor circuits.

To prevent overfusing, a tamper-resistant plug fuse is available. It is constructed in two separate parts as shown in Fig. 1-19. The adapter, similar to the plug, comes in two sizes. One size is for fuses up to 15A, the other for fuses of 16—30A. Due to the construction of the adapter ring and plug it is difficult to overfuse or defeat the fuse in these circuits.

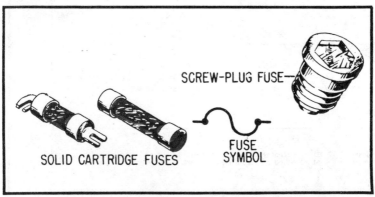

Fig. 1-17. Typical fuses used in home wiring systems.

Fig. 1-18. Time-lag fuses are used in circuits where an electrical device needs a momentarily high starting current.

When a fuse rating larger than 30A is required a cartridge fuse is recommended. The ferrule-contact cartridge fuse is available in ratings up to 60A and is divided into two sizes to prevent overfusing. The two-inch ferrule fuse comes in sizes up to 30A while a three-inch fuse is used for ratings of 31–60A.

In circuits with current ratings above 60A a knife-contact fuse is used. These too come in various lengths to ward against overfusing.

Circuit Breakers

Much more convenient and used nearly exclusively in new construction, the circuit breaker is gradually replacing the fuse as a protective device. Circuit breakers perform the same function as fuses and can be obtained with a time-lag feature similar to slow-blow fuses. There are three types of circuit breakers, each employing a different method of operation.

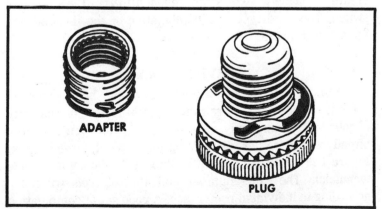

ADAPTER

PLUG

Fig. 1-19. Tamper resistant fuses are used to prevent overfusing.

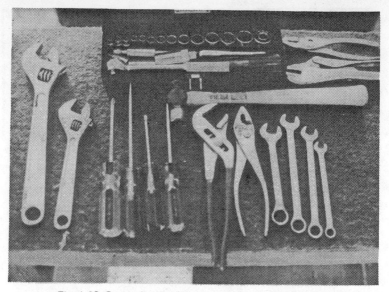

Fig. 1-20. General tools used in washer and dryer repair.

The most common type is the *thermal*. Thermal circuit breakers have a bimetallic element built into the breaker that changes shape as current heats the two metals. Current flowing through the circuit heats these metals. Since they are of different substances they bend at a different rate. The elements act as a latch in the breaker mechanism. When bent they cause the breaker switch to trip.

Magnetic circuit breakers depend on increased current to create a magnetic field strong enough to trip the magnetic switch. The third type is a combination of the thermal and magnetic breakers.

SERVICE TOOLS

Compared to other trades, servicing home laundry equipment does not require many special tools. The majority of jobs will require only basic handtools that you probably already own if you are now doing any type of service work. Figure 1-20 shows the typical toolbox of a laundry appliance technician. The exact number and kind of tools will vary according to individual preference. A good set of combination wrenches (open-end on one end and box-end on the other), a hammer, pliers, screwdrivers, adjustable wrenches, a socket set, a soldering gun, and a voltmeter are the backbone of this particular box. The voltmeter will be discussed in more detail later. First, let's look at some of the special purpose tools you may need if you intend to service many washers and dryers.

Special Purpose

Like any other professional, the home laundry repairman has need of certain special tools not generally used in other work. These tools represent a considerable investment and should not be purchased without first investigating the need and frequency of their use. Some of these tools are used for only one job. Some are only needed when a transmission overhaul is called for, for example. If the tool is only a time saver and you do only a limited amount of this type of work, you should consider carefully the cost of such a tool and the actual amount of time it will save you.

Most special purpose tools (Fig. 1-21) are available from the appliance manufacturer. Some service literature will list the tools you should have in your shop and gives part numbers for these devices. Care should be exercised in the purchase of these tools. If you only intend to service one brand of appliance, your choice should be along the lines of tools specified by that manufacturer. When a complete disassembly of a washer is necessary you will find a spin-basket support (see Fig. 1-21) very useful. These supports vary from brand to brand, however.

Snapring pliers are available in two types and several sizes. One type is for snaprings that must be expanded to be installed or removed. The other type is for snaprings that are compressed for installation and removal. Depending on the make of appliances you service you may need several sizes of these pliers.

Bearing pullers, gear pullers, and spring compressors also come in many sizes and types. Here again the manufacturer's service literature can offer valuable assistance in making the right choice in purchasing your equipment. This is especially true with brakeband tools. The one shown in Fig. 1-21 is used to

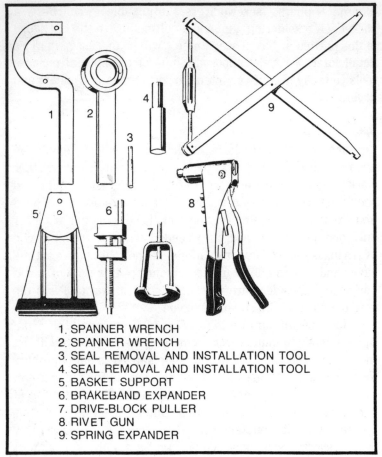

1. SPANNER WRENCH
2. SPANNER WRENCH
3. SEAL REMOVAL AND INSTALLATION TOOL
4. SEAL REMOVAL AND INSTALLATION TOOL
5. BASKET SUPPORT
6. BRAKEBAND EXPANDER
7. DRIVE-BLOCK PULLER
8. RIVET GUN
9. SPRING EXPANDER

Fig. 1-21. Some special purpose tools are needed to repair washers and dryers.

expand the brakeband on one make of automatic washer. This tool may not work on some other makes, however.

It is possible to make some special tools. In fact, you may even design a few of your own to speed up a job or make a certain service more convenient to perform. One example is a special pair of pliers for removing or installing spring-type hose clamps (Corbin clamps). A groove may be cut into each jaw of a conventional pair of slip-joint pliers to make it easier to hold these clamps in an open position. This tool can also be purchased as shown in Fig. 1-22. It makes it much easier to remove or install spring clamps.

44

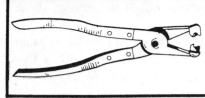

Fig. 1-22. Corbin clamp pliers make it easier to remove and replace spring-type hose clamps.

Electrical Tools

Electrical tools associated with washer and dryer repair are of the general purpose type. That is, they can be used for more than one type of work. Some of these tools can also be made in your own shop.

In working with automatic laundry equipment, you will be making two basic electrical measurements—voltage and continuity. The test lamps shown in Fig. 1-23 can be made with little effort or expense and will do a satisfactory job in most cases. A lamp to check for voltage can be made from an old lamp socket and a pair of test leads. A 120V lamp is connected as shown in the figure. With the appliance connected to the electrical outlet, the test probes can be touched to various points in the circuit to see if voltage is present. If so, the lamp will light.

The other test lamp is used to check continuity in the circuit. Here power must be removed from the appliance and the test lamp is plugged in. The test probes can now be touched

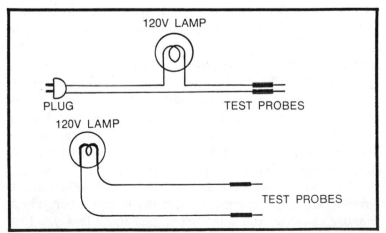

Fig. 1-23. Test lamps are used to make voltage and continuity checks.

to various points to see if current will flow between these points.

A word of caution here. Always use extra care when using test lamps to check a circuit. You are dealing with a live circuit and could easily receive a harmful or even fatal electrical shock.

A good electrical test kit represents a safer and more modern approach to washer and dryer service. For that reason this book will deal only with troubleshooting methods that employ the voltohmmeter (VOM). The meter you select for your toolkit will depend on your own personal preference but it is not necessary to spend a lot of money on an elaborate meter. A meter suitable for appliance repair should have the following features and usually costs less than $25.

- Resistance scales—R × 1; R × 100; R × 10K.
- AC voltage scales—0−2.5V; 0−10V; 0−25V; 0−100V; 0−250V; 0−500V; 0−1000V.
- DC voltage scales—0−0.5V; 0−5V; 0−25V; 0−50V; 0−100V; 0−250V; 0−500V; 0−1000V.
- Direct current scales—0−5 mA; 0−50 mA; 0−500 mA.

Some technicians prefer a meter with an AC feature up to 20−25A. Direct measurements of alternating current are usually much less convenient to make than direct voltage measurements, and very seldom is it necessary to make current measurements. If you find it necessary to make alternating current measurements in your service, a separate ammeter can be purchased for that purpose. This book will show you how to use voltage and resistance measurements to perform almost any type of diagnostic test.

To complete the electrical toolkit, we need to add tools for making electrical connections. Most electrical connections in washers and dryers are made with solderless lugs. A good assortment of lugs and a lug crimper should be a part of this kit. In addition, for those few connections that are soldered, a soldering gun of 100−140W is recommended.

The tools outlined here are the basics—just enough to get you started. As you can see, most of these are contained in a standard toolbox. You will want to add those that you feel make your job easier.

Washer and Dryer Motors

Automatic washers and dryers receive their mechanical power from electric motors. These motors have undergone several changes through the years but have been somewhat standardized in modern laundry equipment. Horsepower varies from one-quarter to one horsepower, depending on the load capacity of the machine. The motor may be single-speed or two-speed and may or may not be reversible. Almost all motors are now split-phase and capacitive-start motors.

These motors usually give little trouble throughout the life of the appliance. Normally, the switches, wiring, and other controls tend to give more problems than the motor itself. Many inexperienced technicians, however, frequently mistake defective motor controls for motor failure.

SPLIT-PHASE MOTORS

Split-phase motors are very simply constructed and consist of only three major parts: the motor frame, the field windings, and the rotor. The frame supports both the field windings and the rotor. Bearings in the endbells of the main frame support the rotor and, at the same time, allow the rotor freedom of motion.

Fig. 2-1. Rotor mechanism of a typical induction motor.

The field windings consist of several coils of wire placed around supporting pole pieces extending from the main frame. Depending on the exact type of motor, there will usually be four or more pole pieces. If the motor is a single-speed type there will be one group of *run* windings, consisting of many turns of heavy wire, and a group of *start* windings, composed of relatively few turns of much lighter wire. Two-speed motors will have two sets of run windings.

A squirrel cage construction rather than windings are used in the construction of the rotor. The rotor consists of a laminated cylinder with slots cut into the surface. The squirrel cage winding usually consists of heavy copper strips pushed into slots cut into the laminated cylinder as shown in Fig. 2-1. Each copper strip is brought out to the end of the cylinder and shorted together by a shorting ring.

The electrical relationship between the squirrel cage rotor and the field windings is very similar to that existing between the windings of a transformer. This relationship can be better understood with the aid of Fig. 2-2. If an alternating current is connected to the primary winding of the transformer, a pulsating magnetic field will be established about the primary winding. This field will expand and collapse with each half cycle of the alternating current. As the current reverses, the magnetic field reverses. The pulsating magnetic field induces a separate current in the windings of the secondary.

If a short is placed across the secondary (as in Fig. 2-2) very little resistance will oppose the induced current. Therefore, this induced current will be high. This current will pulsate as the current in the primary pulsates and will likewise create its own magnetic field. It is the relationship between these magnetic fields that causes the rotor to spin.

Look back at the illustration of the squirrel cage rotor and compare it to the secondary of a transformer. The copper bars in the slots can compare to the windings of transformer secondary. The shorting rings at each end are the shorting jumper at the end of the laminated cylinder. The laminated cylinder serves the same function as the core of a solenoid or transformer: to increase the strength of the magnetic field.

In actual practice the field coils are not wound around individual pole pieces as they are in direct current (DC) or universal motors. They are formed into slots in the frame and arranged to set up a definite magnetic field. It is the interaction of this magnetic field with the magnetic field of the rotor that causes the rotor to spin. The alternating current applied to the field windings goes through its cycle; the magnetic field of the rotor tries to stay up with the moving magnetic field of the field windings. Therefore, rotation is sustained. Note the word *sustained*. Although the run windings can generate the magnetic field necessary to keep the rotor turning, it cannot by itself start the rotation.

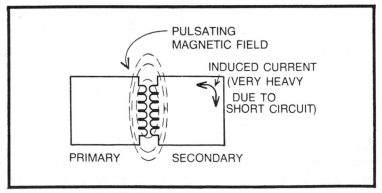

Fig. 2-2. Transformer action in a motor.

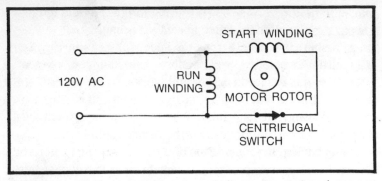

Fig. 2-3. A split-phase motor has two windings: one start and one run winding.

The motors used in most automatic washers and dryers usually employ a method of starting rotation that identifies them as split-phase motors. Either capacitance, inductance, resistance, or a combination of these is used to develop the initial starting torque. A split-phase motor has two windings, one *start* winding and one *run* winding, as shown in Fig. 2-3. With the motor at *rest* no power is applied and the start switch is closed. This places the start and run windings in parallel. Since the start winding is made up of several coils of very fine wire, its resistance is much higher than that of the larger wire with fewer coils. This results in a phase difference in the current flowing through the two windings. This phase difference and the fact that these coils are physically located at different points around the frame cause the magnetic fields generated by the current in the coils to set up a strong magnetic field that will start to spin the rotor. As soon as the rotor reaches its designed run speed the start switch will open and remove current from the start winding. With the motor already running, the run winding will give the rotor sufficient torque to maintain its speed.

Opening and closing of the start switch is controlled by centrifugal force from the spinning rotor. Action of the centrifugal start switch is illustrated in Fig. 2-4. With the motor stopped or running at a slow speed the weights are held inward by the tension of the springs. In this position the start switch is closed and current can flow in the start winding. As the motor achieves the proper run speed, centrifugal force acts on the weights, causing them to move against the tension springs. This causes the collar to move down the shaft toward the rotor. Springs attached to one contact of the start switch will cause it to open and remove the starter winding current. When power is removed and motor speed begins to fall off, centrifugal force on the weights is reduced and the springs return the weights to their rest position. As the weights move toward the shaft, the collar is forced against the contact of the start switch closing the points. The motor is now in its *rest* position waiting for power to be applied and the starting process to begin. The start switch is located inside the motor on older models. In later models the switch contacts have been moved outside the motor shell, with the centrifugal device still located inside. Mechanical motion is applied to the external contacts by a small lever extending from the centrifugal device through an opening in the motor shell.

The starting method most often used with automatic washer and dryer motors is shown in Fig. 2-5. This is known as a split-phase capacitive-start motor. Operation of this motor is exactly the same as the motor shown in Fig. 2-3, except for the

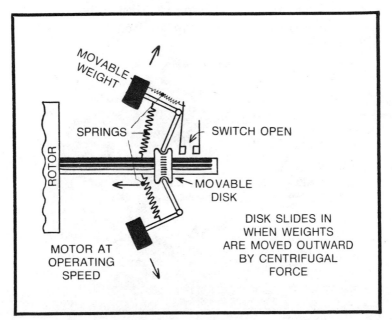

Fig. 2-4. The actions of a centrifugal motor-start switch.

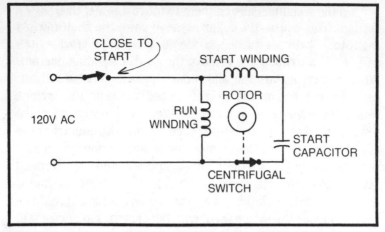

Fig. 2-5. A split-phase motor with a capacitive-start feature. This motor is used most often with automatic washers and dryers.

capacitor action in series with the start winding. Adding the capacitor to the start circuit increases the phase difference between the currents in the two windings, thereby increasing the starting torque. On some appliances, but not too often on laundry appliances, you may find a second capacitor in series with the run winding. Figure 2-6 shows the physical appearance of a capacitive-start motor. Notice the capacitor housing on the motor.

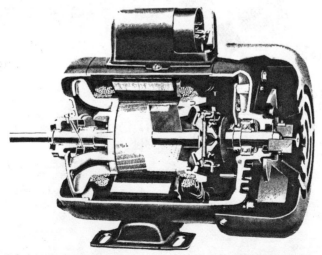

Fig. 2-6. A typical capacitive-start motor. Notice that the capacitor housing is located outside the motor housing.

In some applications it is desirable to be able to reverse the rotation of appliance motors. To do this with split-phase motors it is only necessary to reverse either the start windings or the run windings, but not both. Figure 2-7 illustrates how this is accomplished in some automatic dryers. With 120V AC applied between points 1 and 2 the run winding has the necessary current to sustain rotation. When the motor is at *rest*, current can also flow through the centrifugal switch to the capacitor and the reversing switch. If the reversing switch is in the position shown, terminal 4 at the motor start winding is connected to terminal 3 while terminal 5 is connected to the capacitor. This places the capacitor and start winding in parallel with the run winding. With the motor connected like this it would start to rotate in one direction.

Again, assume the motor to be at *rest* and the reversing switch shifted to the opposite position from the previous example. The start winding is again in parallel with the run winding. This time, however, terminal 4 is connected through the reversing switch to the capacitor, and terminal 5 is connected to terminal 3. In this position the start winding is

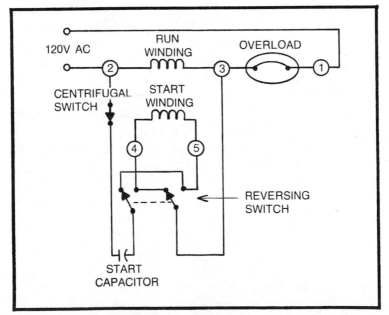

Fig. 2-7. Reversing a capacitive-start motor.

exactly reversed from its previous position. The motor will start as it did before but the direction of rotation is reversed.

WASHER AND DRYER MOTORS

The general features of the various motors we have been discussing are found in most washer and dryer motors. There are, however, a few mechanical and electrical characteristics with which you should be familiar.

Mechanical Features

There are two basic types of construction used in washer and dryer motors, one type illustrated in Fig. 2-6. The mechanical features of this type allow the motor to be completely disassembled for repairs. The amount of repairs you perform inside the motor shell will depend on your own service operation. Bearing replacement and repairs to the centrifugal device and switch are usually made by appliance service technicians. Major repairs to the motor, such as rewinding, are the job of a service shop.

Four long bolts extending from endbell to endbell through the frame normally hold the motor together. Before removing these bolts, scribe the relative location of the endbells to the main frame with a small chisel. Center the chisel across the joint between the endbell and frame. Tap the chisel with a hammer. Repeat this at the opposite end of the motor. Be very careful to note the exact order in which parts are removed to insure easy and accurate reassembly.

Once the bolts are removed, the endbells may be removed for access to the internal parts of the motor. Be careful not to damage the centrifugal switch when removing the endbells. If an endbell is particularly hard to remove, it may be tapped with a rawhide or rubber-faced mallet to loosen it. Do not strike the endbells with a steel-faced hammer since they may crack.

Most motors are mounted in the appliance with various types of flexible mounting. One type holds the motor in place by a clip-type mounting spring that straddles a tight fitting rubber ring over the endbell.

Frequently, the T-frame motor is referred to as a throwaway motor because it is not designed to be repaired by the service technician. Instead, once the technician has determined that the motor is definitely at fault the entire motor is replaced.

T-frame motors are built with the same general mechanical features as the earlier motors except that the endbells are permanently attached to the motor frame with a special bonding material. The centrifugal start mechanism is located inside one of the endbells, and the mechanical motion of the device is transmitted to the centrifugal switch by a small lever extending through a hole in the endbell as mentioned earlier. These motors are not to be disassembled in the field. If mechanical problems or electrical troubles inside the motor housing are encountered, the entire motor should be replaced. In case of electrical troubles always make certain the problem is in the motor and not the switch.

Electrical Features

There are three types of motors used in modern home laundry equipment (classified according to windings and other electrical features): single-speed, two-speed, and reversible. Single-speed motors usually have a four-pole winding arrangement and rotate at about 1700 revolutions per minute (RPM), while two-speed motors have either a four-pole or six-pole winding arrangement and rotate at approximately 1700 RPM and 1100 RPM, respectively.

Usually, T-frame motors include an overload protector inside the motor housing. This device is in series with the windings of the motor and reacts to excessive temperature or current. If either of these conditions exist the protector opens the circuit through the motor windings and stops the motor. Once the overload condition has been removed the protector will close the circuit through the windings after a cooling period of 5 to 10 minutes. Since the overload protector is inside the housing of a T-frame motor, the entire motor must be replaced if the protector becomes defective. Starting and reversing of these motors is controlled by the motor-start switch and other external controls.

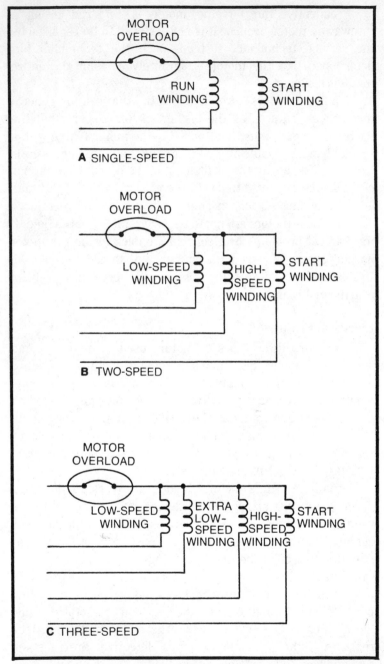

Fig. 2-8. Wiring diagram of three motors: (A) single-speed; (B) two-speed: (C) three-speed.

Internal electrical features for most laundry appliance motors are shown in Fig. 2-8. Figure 2-8A is for single-speed motors. As shown in the schematic, there are two separate windings. One side of each winding is connected to one side of the 120V AC circuit through the built-in overload protector. Two-speed motors are practically identical except for the low-speed winding (Fig. 2-8B). The speed at which this motor runs is determined by the switch selecting either the low- or high-speed windings.

Shown in Fig. 2-8C is a three-speed motor. Although not as common as the single- or two-speed motor, this motor will be found on some automatic washers. It is very similar to the two-speed motor except for a third group of pole windings which give the motor an extra-slow speed of approximately 800 RPM.

MOTOR CIRCUITS

To test motors properly the service technician must fully understand how the motor works under normal conditions. This is especially important with throwaway motors because of the replacement cost. When determining if the motor must be replaced, first study its operation while installed in the machine. After removing the motor, test it on the bench to make sure you have made a correct judgment. Always doublecheck your test results to avoid unnecessary replacements.

Single-Speed Operation

The single-speed nonreversing motor circuit is shown in Fig. 2-9. Three leads, labeled C, R, and S in the diagram, extend from the motor and connect it to the external circuit. The common wire (C) is connected to the neutral leg of the 120V AC circuit. Notice that the common line is connected in series with both the start and run windings so as to break the circuit to both windings under overtemperature or overcurrent conditions. Run winding R is connected directly to the start/stop switch shown in the *stop* position.

A starting capacitor is connected to the start winding, lead S. The opposite side of the start capacitor is connected to a

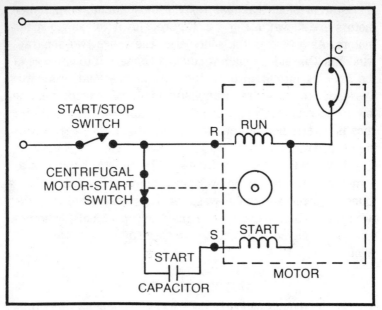

Fig. 2-9. Single-speed motor circuit (nonreversing).

terminal on the motor-start switch. With the motor at *rest*, the start switch connects the capacitor to the start/stop switch as shown. Careful examination of the diagram will show that the start capacitor and start winding are now in parallel with the run winding.

At the moment the start/stop switch is closed, current flows through both the start and run circuits and the series-connected overload protector. With current present in both windings, the motor will begin to turn. When the motor reaches its approximate operating speed, mechanical motion (shown by the dotted line in the figure) from the centrifugal device will cause the motor-start switch to open. This action removes current from the start winding and allows the run winding to continue to supply sufficient torque to keep the motor rotating at its proper speed.

The major difference between the single-speed nonreversing motor in Fig. 2-9 and the single-speed reversing motor in Fig. 2-10 is the manner in which the run and start windings are connected to the external circuit. Notice (in Fig. 2-10) how the run winding is connected to the external circuit.

Current is supplied from the 120V common through the overload, the run winding, and terminal R1 to the start/stop switch. At point R2, the connection between the overload and the run winding, a lead is run to terminal 1 of the reversing switch. The hot side of the 120V supply is connected from terminal R1 through the motor-start switch and start capacitor to terminal 2 of the reversing switch.

With the reversing switch in the position shown, terminals 1 and 4 are made and terminals 2 and 6 are made. Therefore, S1 is connected to R2, and S2 is connected to R1. Since the motor is at *rest*, the motor-start switch has connected the start winding in parallel with the run winding. Closing the start/stop switch will cause current to flow through both windings in

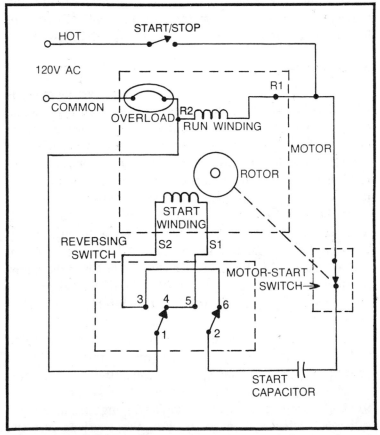

Fig. 2-10. Single-speed motor circuit (reversing).

parallel and the motor will turn. At the designed speed the centrifugal device opens the motor-start switch and the motor continues to rotate in whichever direction it was started.

Assume once again the motor is at *rest* with the start/stop switch open and the motor-start switch closed as shown in the figure. However, this time let's assume the reversing switch is placed in the opposite position, causing terminal 1 of the switch to connect to terminal 3 and terminal 2 to connect to terminal 5. Again the start winding is in parallel with the run winding but S1 is connected (through the capacitor and motor-start switch) to R1; S2 is connected to R2. This is reversed polarity from the preceding example. Under these conditions, if the start/stop switch is closed the motor will start but in the opposite direction. The single-speed nonreversing motor and the single-speed reversing motor operate identically, except for the initial starting action caused by the reversing switch.

Two-Speed Operation

A simplified diagram of a two-speed nonreversing motor is shown in Fig. 2-11. Note the similarities between Fig. 2-9 and Fig. 2-11. The most noticeable difference is the second set of contacts in the motor-start switch. Although it is not a universal practice, many multispeed motors are wired so that, regardless of the operating speed, the motor always starts as if it were to run on high speed. Then, if low speed is selected, when the operating speed is reached, the motor-start switch will cause the current to be shifted to the low-speed winding. This is done in order to take advantage of the higher initial torque generated by the high-speed windings. Since the motor reaches its operating speed in the first few seconds of operation, it is very difficult to notice the shift from high to low speed.

In Fig. 2-11 the neutral line of the 120V supply is routed through the upper contacts of the motor-start switch and capacitor to the start winding, and through the speed-selector switch and motor-start switch to either the high- or low-speed windings. Once again all of the motor windings are in parallel with each other and in series with the overload and the start/stop switch. Following the wiring diagram in Fig. 2-11, assume the speed-selector switch is in the position shown. Current flows through both the start and high-speed windings. Since the motor-start switch holds the circuit to the low-speed winding open, no current can flow in that circuit. As proper speed is reached the motor-start switch opens the circuit to the start winding. At the same time the lower set of contacts shifts from the high-speed winding to the low-speed winding. However, since the speed-selector switch is made between contacts 1 and 3, and terminal 2 is open, no current can flow in the low-speed winding. The high-speed winding continues to supply the torque.

Now let's back up to the original conditions with the motor stopped. Shift the speed-selector switch from *high* to *low* speed. In this instance, contacts 1 and 2 are closed and contact 3 is open. When the start/stop switch is closed, the motor will start on the run and high-speed windings just as it did before.

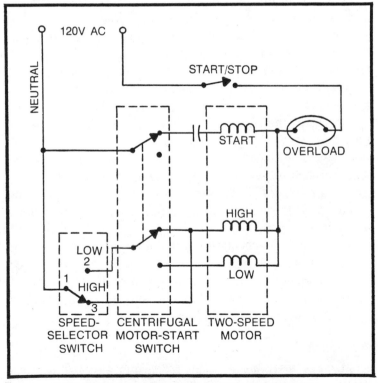

Fig. 2-11. Two-speed motor circuit (nonreversing).

However, this time, when the centrifugal start switch changes position, terminal 2 of the speed-selector switch is connected to the low-speed winding. This removes current from the high-speed winding and allows the low-speed winding to take control.

Three-Speed Operation

Three-speed motors are identical in operation to their two-speed counterparts, with the single exception of an additional *run* winding to obtain the third speed. The addition of a third run winding to Fig. 2-11 would make that circuit illustrate the action of a three-speed reversible motor.

MOTOR SERVICING

Motor servicing is an important part of repairing laundry appliances. This servicing includes all the procedures necessary to locate and correct any malfunction in the motor or its associated circuits. Also included is spotting possible troubles and correcting them before major trouble develops. The importance of proper troubleshooting is illustrated by what a service representative of a leading manufacturer of automatic washers and dryers told me. He stated that motors are unnecessarily replaced more often than any other component of their machines. This was verified by checks made on *in warranty* motors returned to the manufacturer for replacement. Although some of these replacements may be made through honest mistakes, far more are made by poor servicing practices. In this topic we have attempted to show how to avoid these unnecessary replacements through good servicing procedures.

All troubles encountered with electric motors used in laundry appliances can be traced to either mechanical or electrical problems. The first step then, after determining that the fault is associated with the motor, is to determine whether or not the trouble is mechanical or electrical.

Mechanical

Mechanical troubles range from minor problems that cause a noise, which may be harmless but annoying, to a frozen mechanical shaft that completely locks the motor. Customer complaints about mechanical problems may include any of the following: noisy operation, motor overheats, motor doesn't run, appliance hums but doesn't run, occasional failure to start, appliance blows fuses, or just about anything else you can imagine. In a later chapter we will show you how to determine if the problem is caused by the motor or some other part of the appliance. Here we will look at methods of locating and correcting troubles once it has been determined that the motor, or its associated components, is the cause of the problem.

To verify that the trouble is in the motor, it should be disconnected from its load. This can usually be done by simply removing the drive belt from the pulley. In some cases the motor shaft extends from both ends of the motor and a belt is attached at each end. Once the load has been removed from the motor, you should check the freedom of the motor shaft by rotating it with your hands. The shaft should rotate freely. Now check shaft endplay by pushing and pulling the shaft back and forth toward the motor. There should be only a slight amount of endplay. Insufficient endplay will cause binding, while too much endplay can cause binding, noisy operation, or even failure to start.

Complaints of noisy operation can sometimes be traced to the motor mounting. Look for loose or weak springs in the mounting and hardened or deteriorated rubber mountings that will cause motor vibration or binding. In this instance you would also want to check the condition of the belts used to drive the load and the pulleys. Noise originating in the motor itself will usually be caused by excessive endplay or defective motor bearings. To correct either of these problems, the motor will have to be disassembled. If the motor is the T-frame variety it must be replaced. In motors that can be disassembled, defective bearings can be replaced and excessive endplay sometimes eliminated by installing spacers.

Mechanical checks of the motor itself are limited to these simple operations. If the mounting, pulleys, and belts are in good shape, the rotor is free to rotate, and there is correct endplay, the motor is mechanically in operating condition. We have not eliminated electrical troubles, however.

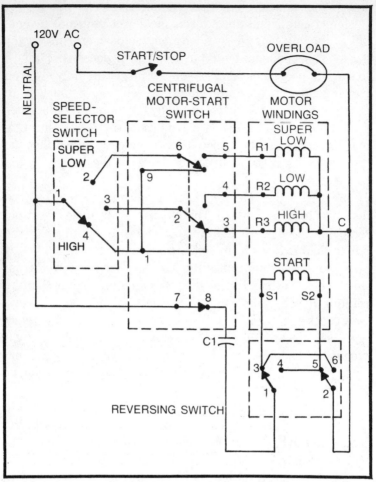

Fig. 2-12. Two-speed motor circuit (reversing).

Electrical

If electrical troubles with the motor are suspected, the motor should be tested while in the circuit. Most electrical problems are traced to the associated circuits rather than the motor itself. For this reason it is imperative that these circuits be thoroughly tested before a motor is replaced. A three-speed reversible motor is used in Fig. 2-12 to represent the manufacturer's service literature. You should first study this literature to familiarize yourself with the operation of the appliance. Determine just what the circuit is supposed to do and verify the operation by running the motor under all possible conditions.

The motor illustrated in Fig. 2-12 is wired for three different rotor speeds in either direction. Therefore, there are six separate modes of operation for this motor. This can be seen by closer examination of the components in the figure. The reversing switch can be set to parallel the start winding with the run windings to cause clockwise or counterclockwise rotation, depending on the polarity of the connection. Notice in the figure that terminal 1 of the reversing switch is connected through the capacitor and contacts 7 and 8 of the motor-start switch to the neutral leg of the 120V supply. The individual run windings are also tied to the neutral leg through the motor-start switch and the speed-selector switch. Terminal 2 of the reversing switch is connected directly to the opposite end of the run windings at the junction common with the overload protector. Voltage is supplied to these windings through the protector and the start/stop switch.

With the reversing switch in the position shown, contacts 1 and 3 are closed and contacts 2 and 5 are closed. This connects S2 of the start winding to the common connection of the run windings and the overload protector. S1 of the start winding is connected through the capacitor and motor-start switch to terminal 1 of the speed-selector switch. The run windings are connected to the same point through the centrifugal motor-start switch and the contacts of the speed-selector switch.

Placing the reversing switch in the opposite position causes terminal 1 of the reversing switch to contact terminal 4 and terminal 2 to contact terminal 6. Under this condition, S1 is connected to the common connection of the run windings and the overload protector. S2 is now connected through the capacitor and motor-start switch to the opposite ends of the run windings at terminal 1 of the speed-selector switch. This reverses the polarity of the start winding with respect to the run windings and causes the motor to start in the opposite direction.

The neutral side of the 120V supply is connected to one of the run windings through the speed-selector switch and motor-start switch. When the speed-selector switch is in the *high speed* position, contacts 1 to 4 are connected to supply voltage to terminal 1 of the motor-start switch. In the *low speed* position, contacts 1 to 3 of the speed-selector switch are connected and voltage is supplied to terminal 2 of the motor-start switch. In the *superlow* position, contacts 1 and 2 of the speed-selector switch are made and voltage is supplied to terminal 6 of the motor-start switch.

With *high* speed (terminal 4) selected and the start/stop switch closed, current flows through the high-speed winding and contacts 1 and 3 of the motor-start switch. When the motor reaches the proper operating speed the centrifugal switch opens contacts 2 and 3 and contacts 7 and 8. This removes current from the start winding while current flow remains in the high-speed winding due to the connections between 1 and 3 of the centrifugal switch.

If *low* speed is selected at the speed-selector switch, then contacts 1 and 3 of that switch are closed. Now when the start/stop switch is closed, current flows through these contacts and contacts 2 and 3 of the motor-start switch. As the motor reaches operating speed, the centrifugal switch opens contacts 2 and 3 and opens 2 and 4. Contacts 7 and 8 open at the same time and the motor runs in the direction selected by the reversing switch at slow speed.

Selecting *superlow* speed causes terminals 1 and 2 of the speed-selector switch to make contact. This connects the neutral line to terminal 6 of the motor-start switch. When the motor is stopped the neutral leg is connected from terminal 6 of the motor-start switch through contacts to terminal 9. A jumper from terminal 9 to terminal 1 connects the high-speed run winding to the neutral leg. As the motor picks up speed the motor-start switch opens contacts 6 and 9 and contacts 7 and 8. This removes current from the start winding and the high-speed winding. This leaves only the superlow-speed winding, connected through 5 and 6 of the motor switch, to the neutral leg.

If you can understand the operation of these circuits, you should have no problem with the majority of those found in automatic laundry equipment. Most are simpler than the one presented here.

We said earlier that troubleshooting the motor and its circuits begins with observing machine operations. Using the circuit just explained, you can see that you must operate the appliance in all six modes of operation. First, select a direction of rotation. Then attempt to operate the appliance in each of the three available speeds. After this, try the opposite direction of rotation and all three speeds. Failure of the appliance to operate properly in any mode will require further investigation.

After observing the operation of the appliance, visually inspect the circuits to locate any electrical trouble. Disconnect the appliance from the power source and remove the necessary covers to inspect the components associated with the motor and its wiring. Look for broken leads, shorted or bare wires, and loose connections. Repair these before going any further. Carefully inspect the mechanical connection between the motor and the centrifugal motor-start switch on motors with external switches. A loose mounting can prevent this switch from operating properly.

During the operational checks make note of which modes seem to be malfunctioning. This will speed the electrical checks made with a voltmeter or ohmmeter. Some technicians still use test lamps to check the circuitry; however, a voltmeter is easier and safer to use. An ohmmeter is safer still, since a mere 3V potential is used to test the circuits. This book deals exclusively with the safer voltohmmeter (VOM) checks, and they are the only ones recommended here. First let's look at voltage checks.

Since voltage checks are made with power applied to the appliance, care should be exercised to avoid shock to the serviceman or damage to the appliance.

To check the operation of the start/stop switch with power applied to the appliance, set an AC voltmeter to a scale that will read 120V accurately. Connect one probe to the *neutral* line in Fig. 2-12. Touch the other probe to terminal 2 of the start/stop switch. With the switch open, no reading should be obtained on the meter. When the switch is closed (appliance started) the meter should read about 120V.

Check the speed-selector switch in a similar fashion. Connect one probe of the meter to terminal 1 of the start/stop switch. Select high-speed operation at the speed-selector switch and touch the other probe to terminal 4. The meter should read about 120V. Move the switch to the *low speed* position and measure the voltage at terminal 3. Again expect a reading of 120V. The *superlow* speed position should result in a similar voltage at terminal 2.

Testing the reversing switch can be done with a voltmeter connected between S1 and S2 of the motor-start winding. The reversing action of the switch cannot be checked in this manner if voltage is present when attempts are made to start the motor. Reversing action *can* be checked by observing the rotation of the motor shaft. The voltage at S1 and S2 will only be present during the first few seconds of motor starting. Once the motor is up to speed, contacts 7 and 8 of the motor-start switch are open and the voltage is removed from S1 and S2. Absence of a voltage at these points during the starting phase can mean either the reversing switch or the motor-start switch (contacts 7 and 8) is defective. To isolate such a problem, connect the voltmeter between C on the motor and terminal 8 of the motor-start switch. No voltage here during the first moments of starting points to a defective motor-start switch.

Other contacts of the motor-start switch can be checked by connecting one probe of the voltmeter to terminal C, common to the motor and the run-winding contacts of the motor-start switch. To check the high-speed contacts, connect the second probe to terminal 3 of the motor-start switch and place the speed-selector switch in the *high speed* position. When the motor starts, 120V should be indicated by the meter. Also, in both low-speed settings voltage should be present during the starting moments of operation but should be removed once proper speed is obtained. In low-speed operation, voltage should be present between terminal 4 of the motor-start switch and terminal C of the motor—between terminals C and 5 for superlow speed.

Voltage checks can be used to check the motor itself. If the motor fails to start, check the voltage across the start windings and the high-speed run winding. With voltage present at both these points, and the motor not overloaded, the motor should start. If not, the motor should be removed and bench tested.

In the circuit shown in Fig. 2-12, if one of the low-speed windings is open the motor will start, achieve speed, and shift to the selected low-speed winding. The speed will drop since the open winding cannot sustain rotation. As the motor slows, the motor-start switch will return to the *start* position and pick up speed once again. This can be detected by intermittent voltage across the selected low-speed winding.

Resistance readings taken with an ohmmeter can be used to duplicate many of the tests performed in the preceding illustrations. Since these tests are made with power removed from the equipment under test, they offer the safest ways of testing electrical components in laundry appliances.

Testing the motor-start switch is very simple. With the appliance unplugged, connect an ohmmeter (set to a low range) between terminals 1 and 2 of the start/stop switch. As the switch is opened and closed the meter should alternately read infinite and zero resistance. If so, the switch is probably working as it should. The speed-selector switch may be tested in a similar manner. Connect one lead of the ohmmeter to terminal 1 of the speed-selector switch and the other to terminal 4 of the same switch. Place the speed-selector switch in the *high* position. The meter should read zero resistance. Move the switch to either of the low-speed positions. An infinite reading should be obtained. Move the probe from terminal 4 to terminal 3. Select the *low speed* setting and observe the meter reading. It should be zero. Move the speed-selector switch to the *superlow* position and the meter reading should change to infinite. Remove the probe from terminal 3 and touch it to terminal 2. Again the reading should be zero; when the switch is moved to any other position, infinite.

To make complete tests of the motor-start switch using an ohmmeter, you must be able to move the switch position from *start* to *run*. This can be done by locating the mechanical connection between the centrifugal device and the switch. Close observation will usually show this connection can be moved using a small screwdriver for a lever. Care should be exercised to prevent damage to the centrifugal device or the switch itself. Connect the ohmmeter between terminals 7 and 8 of the switch. When the motor is at *rest* the meter should read zero resistance. If the switch is manually moved to the *run* position the meter should read infinite resistance. Similar tests can be made between the other contacts of the motor-start switch.

Resistance readings can also be used to check the reversing switch. First check the resistance between terminals 3 and 6 and between terminals 4 and 5. In both cases the meter should read zero. Now place the reversing switch in either position and measure the resistance between terminals 1 and 3. If the meter reads zero, the reading between terminals 2 and 5 should also read zero. Terminals 1 and 4 and terminals 2 and 6 should read infinite. Reversing the position of the motor-start switch should also reverse the readings.

The capacitor can be checked with an ohmmeter set to a scale of R×100Ω. Use the shank of a screwdriver with an insulated handle to short the terminals of the capacitor to remove any residual charge. Connect the ohmmeter to the capacitor terminals. The meter should momentarily deflect toward zero and gradually move to a higher reading as the capacitor takes on the charge of the meter batteries. If this happens, the capacitor is probably okay. If the meter reads zero resistance, the capacitor is shorted. If an infinite resistance is obtained, the capacitor is open. Should the meter read any resistance without the needle deflecting toward zero and returning to the reading, the capacitor is leaky and should be replaced.

A word of caution about replacing motor capacitors. Although the capacitors look physically similar to the electrolytic capacitors used in TV and radio circuits, they are electrically different. Always use a motor capacitor of proper value when making a replacement. Never substitute a TV electrolytic capacitor.

Ohmmeter readings are also useful when testing the motor itself. Measuring the resistance of each winding with an ohmmeter can reveal open windings. If all windings read open (infinite), the overload protector is likely open. If the motor has been run or has had power applied during the last few minutes, give the motor 5 to 10 minutes to cool. Then check the resistance readings again. If the circuit is still open, the motor is defective and should be replaced. When making any resistance reading, make sure there is nothing in parallel with the circuit under test that might give false indications on the meter.

Don't overlook the possibility of an open lead in the wiring when troubleshooting for defects in motor circuits. Resistance readings and visual checks can be used to locate broken connections or leads.

The circuit shown in Fig. 2-12 is not meant to represent any particular make or model machine. It is a typical motor circuit similar to many used in automatic home laundry equipment. Using the manufacturer's servicing diagram, the tests used with this circuit can be adapted to almost any motor-driven laundry appliance.

BENCH TESTING AND SERVICING

Once you have determined the motor itself is faulty, it should be removed from the machine for bench testing. There's only a limited amount of bench service that can be done on a T-frame throwaway unit. Aside from cleaning lint from the openings in the frame, replacing rubber mounting pieces, and testing the electrical condition of the motor, little else can be done. Older motors, however, can be opened for service. For that reason we will use an older motor as our example here for bench service.

Operational Tests

The motor should be removed from the machine and tested on the bench for proper operation. During these tests, there will be exposed current-carrying wires and terminals. Be careful not to come into contact with these circuits and prevent any metal from touching them. It is good practice to have a special workbench on which operational tests are made. The top of the bench should be covered with wood, rubber, or insulative tile to prevent accidental electrical shocks.

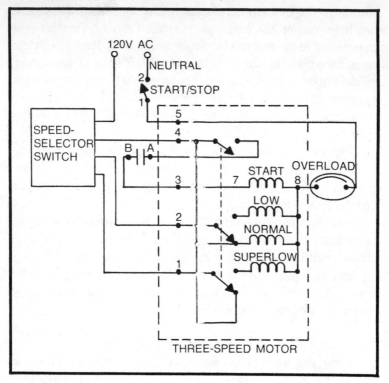

Fig. 2-13. Three-speed nonreversing motor circuit.

In order to demonstrate motor testing procedures we will make reference to a specific motor. For this purpose we have chosen a three-speed motor from a popular automatic washer. This particular motor was chosen for several reasons. As can be seen in the photograph in Fig. 2-6, the starting capacitor is attached to the motor frame and wired internally to the motor wiring and centrifugal motor-start switch. Internal wiring connections are shown in the schematic in Fig. 2-13. External voltages applied to the four input terminals control motor speed. The neutral terminal is connected through control switches to the neutral leg of the 120V AC supply whenever the motor is to be started. Selecting normal operating speed connects the normal terminal to the hot side of the 120V line. Selection of *low* or *superlow* would connect the proper terminal to the hot side. Operation is the same as that explained in conjunction with Fig. 2-12.

Before making any electrical tests, check the general mechanical condition of the motor once again. Rotate the shaft to make sure the rotor is free and not bound by frozen bearings or a bent cooling fin on the rotor. Check for endplay by pushing and pulling on the shaft. You are now ready for the electrical checks.

Be on the safe side. Start your electrical tests with a complete set of resistance measurements. Set the ohmmeter on the highest resistance scale and measure the resistances between each motor terminal and the motor frame. Any resistance here indicates an electrical leakage between the motor windings and the motor frame. This usually means a breakdown in the winding insulation and, in most cases, will require the services of a motor shop unless you do your own rewinding. The permissible amount of leakage will depend on the particular motor and is sometimes specified in the manufacturer's instructions. Excessive leakage results in poor efficiency, sometimes causing the motor to run hot. On ungrounded appliances this condition can cause electrical shocks.

Measurement of the resistance at the motor terminals is the next step. Exact readings are difficult to predict since the

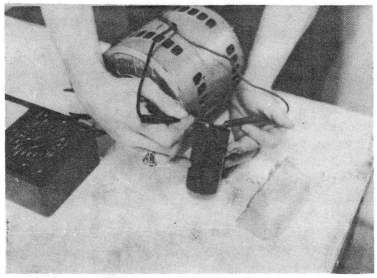

Fig. 2-14. Capacitor testing with an ohmmeter.

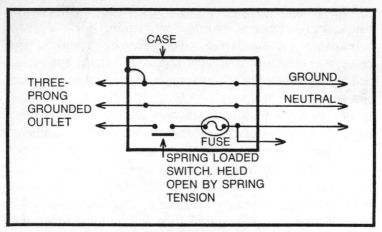

Fig. 2-15. A wiring diagram of a test box.

number of turns in the windings will vary with the motor type and manufacturer. These readings will mean more to you as you become familiar with various types of motors and motor-switch arrangements. We must remove the capacitor cover to gain access to the capacitor terminals (Fig. 2-6). Once the cover is removed (Fig. 2-14), one lead should be unsoldered from the capacitor to prevent the parallel resistance of the windings from causing erroneous readings. Touch one meter probe to each of the capacitor terminals. The needle should swing toward zero then slowly climb. Some resistance, indicating capacitor leakage, will be measured but this should be very high.

Connect one probe of the ohmmeter to the normal terminal and alternately touch the other meter probe to each of the capacitor leads. (*Note*: One lead is still disconnected. Make sure you touch this lead and *not* the terminal from which it was removed.) Take note of the two readings. One will be very low, only one or two ohms at the highest, while the other lead may read very high, but not infinite. The near-zero reading indicates you have located the capacitor lead that connects with the motor-start switch. This also means the contacts of the motor-start switch are closed, as they should be since the motor has no power applied. You should have a high reading between terminal B of the capacitor and the *normal* terminal of the motor (Fig. 2-13). since meter current must flow through both the start windings and the normal windings to get from probe to probe.

Leave one probe of the meter connected to lead B and move the other to the *neutral* terminal. You should have continuity—not infinity but a lower reading than you had before, since only the start winding is in the circuit. An infinite reading means an open exists between lead B and the *neutral* terminal—probably an open overload. Zero resistance (seldom encountered) would mean a short across the start winding. If the motor-start switch is operating properly you should have almost zero resistance between any of the three speed-control leads. Use the wiring diagram for the motor with which you are working and check the various windings and switch contacts.

If the motor does not check out at this point there is not much use in actually running the motor. Instead, any defects noted in the resistance or mechanical qualities should be corrected. Once the motor passes these tests, however, it is time to run the motor under its own power with the capacitor reconnected. This can be done with the test box shown schematically in Fig. 2-15. Test boxes like this can be purchased or shop-made. Power is supplied to the test box through a conventional three-prong grounded plug. The ground wire is tied to the case of the test box and to another lead going to the motor. The neutral wire is brought straight through the test box to a second motor lead. A spring-loaded momentary-contact switch and fuse (no more than 30A) is placed in series with the hot side of the line. Two hot leads are provided to make the test box adaptable to almost any motor wiring scheme.

To use the test box with a one-speed internal motor-start switch, connect the box to the motor as shown in Fig. 2-16. Connect the ground lead to the motor frame (F), the neutral to the motor common (C), and one of the hot leads to the run-winding terminal (R). The second hot lead is to one terminal of the capacitor. Attach a jumper between the second lead of the capacitor and the start-winding terminal (S) on the motor. Either block the motor securely on the bench or have an assistant hold the motor securely in place with insulated

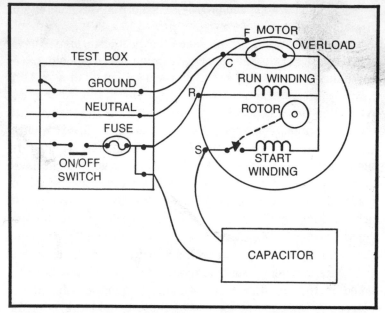

Fig. 2-16. A wiring diagram showing the test hookup for a single-speed motor.

gloves during the remaining part of these tests. Plug the power cord into a conventional three-prong grounded outlet. To test the motor it is only necessary to press and hold the test switch.

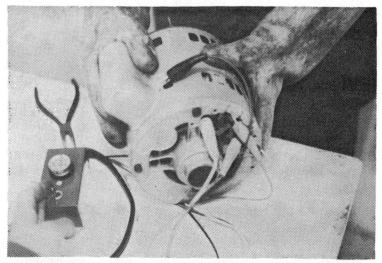

Fig. 2-17. A motor being tested with a test box.

This same test setup can be used to test the motor capacitor. Should you suspect the capacitor as a trouble source, a substitute can be used for the test. Figure 2-17 is a photograph of a motor being bench tested with a test box.

Interior Checks

When servicing motors that have removable endbells, a certain amount of service and repair work can be profitably undertaken by the appliance technician. Bearing replacements and repairs to the centrifugal motor-start switch are typical. Figure 2-18 shows the first step in the disassembly of these motors. A light chisel and a small hammer can be used to scribe index marks on the endbells to insure they are properly aligned when the motor is reassembled. After the endbells have been scribed, the four bolts holding the endbells to the main frame can be removed.

Gently tap the endbells with a wood or rawhide mallet until they can be separated from the frame by hand. Once the endbells are removed, the centrifugal mechanism and switch contacts will be exposed as shown in Fig. 2-19. Check to see that the centrifugal switch will operate smoothly by pressing on the trip mechanism. Switch contacts should be clean and free of pits and burn marks.

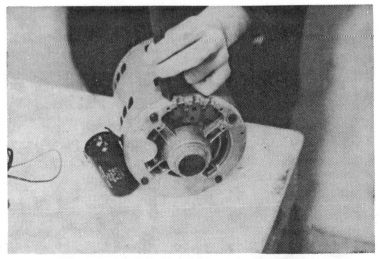

Fig. 2-18. The first step of motor disassembly.

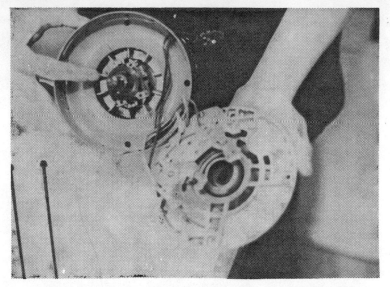

Fig. 2-19. The centrifugal switch mechanism in a dismantled motor.

Bearings used in these motors are usually of the bushing type and pressed into the endbells. They can usually be removed and replaced by tapping them with a hammer. Exercise care not to damage endbells or bearing surfaces during this operation. In extreme cases it may be necessary to remove these bearings with a bearing press (the type found in electrical motor shops).

When reassembling the motor, take care not to damage the switch contacts or pinch a wire between the endbell and the motor frame.

Chapter 3

Automatic Washer Installation

This chapter is devoted to the proper installation of automatic washers. Although most of your work will be concerned with machines that are already installed, you will find that faulty operation can often be traced to improper installation. Some independent technicians also contract the delivery and installation of laundry appliances for stores not having a service department. This can be a good source of additional income for the small repair shop. Space does not permit installation details for each model of every manufacturer. The general requirements for most machines are given, however, and details can be found in the manufacturer's service literature. From the information given here you should be able to recognize an adequate washer hookup. Make a habit of checking the hookup each time you make a service call. The added sales of hoses, valves, filter screens, and other small parts will pay for the small amount of time it takes to look over the installation.

REQUIREMENTS

The automatic washer must have an adequate supply of both cold and hot water to do the job it was designed to do. Poor water pressure will increase the time required for the washer to complete its cycle. Therefore, proper pressure should be maintained. Water containing sand and other debris can damage not only clothes but the machine as well. A drainage system large enough to handle the water discharge must be provided for removing used water. The water supply must also be protected from freezing.

The weight of the machine itself, combined with the weight of the clothes and water in the machine, requires a solid, level floor for support. If the floor is unsteady, the machine will vibrate and "walk" around. An unlevel floor will result in improper water levels.

Most machines require a 15A electrical service. If at all possible this should be a separate circuit protected by a 15A breaker. A three-prong grounded receptacle should be used. In addition to the ground provided by this circuit, it is a good safety practice to have an additional ground between the metal cabinet and the cold-water pipe (if metal pipe is used).

When installing new machines your customer may ask for advice concerning the best location. Convenience to the homeowner should be considered second to operational requirements. Storage area for laundry products and accessories must also be considered. Be sure to talk this over carefully with the homeowner and, when possible, tailor the installation to fit those needs and desires.

LOCATION

Many new homes have specially built laundry rooms with the necessary storage, water, and electrical facilities provided. In this case you will be more concerned that the washer will fit into the specified location. I once knew a very disappointed housewife who discovered that the lid of her new washer would not open without striking the overhead built-in cabinets. Most manufacturers include a chart in the literature accompanying new appliances. These charts show the minimum clearances required for proper installation. If you work with a sales organization, make sure that the sales personnel are aware of these specifications and discuss them with the customer.

Once the location is selected and the washer is delivered, it must be set up. Removing the washer from the shipping crate and disconnecting the shipping braces are usually done at the location, and debris from this uncrating is removed from the home. Packaging and bracing will vary from make to make but usually there are clamps and pads that hold the tub and its suspension mechanism firmly in place during shipment.

The machine will require leveling at the spot where it will sit. Some models have an automatic leveling feature that will maintain a level position without further adjustment once initial setup has been made. Other machines have individually adjusted legs that are varied in height to the desired level. If adjustable legs are provided, it is best not to extend them more than 1¾ in. The longer the legs are extended the less secure the machine will sit.

Figure 3-1 illustrates how the adjustable legs should be installed. In most cases a bolt is welded to the base of the machine. The legs are screwed into this bolt, and a locknut is tightened to hold the leg in place. In some cases a spring clip is attached to the legs for shipment. This clip should be removed before attempting to level the machine. Rubber foot pads should also be installed under the legs as shown.

A spirit level can be used to determine the correct level. It is best to make two level measurements on the top of the

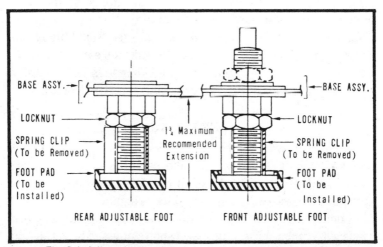

Fig. 3-1. Adjustable leveling screws on an automatic washer.

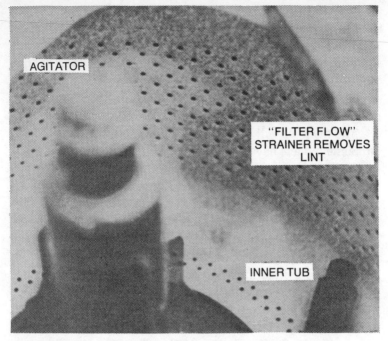

AGITATOR

"FILTER FLOW" STRAINER REMOVES LINT

INNER TUB

Fig. 3-2. Judging machine leveling by observing the waterline.

machine. Take one measurement across the width of the machine near the lid and another across the depth. After hookup let the machine fill with water and notice the relative distance between the waterline and the holes in the spin basket. The waterline should be approximately the same distance from all the holes at that height (see Fig. 3-2). Try to shake the machine with your hands to make sure it is steady. It is very important that locknuts on all four legs are securely tightened after the washer is leveled and in its final location.

WATER HOOKUP

Automatic washers require an adequate supply of both hot and cold water. The hookup should be equipped with shutoff valves for both supplies. If at all possible the shutoff valves should be accessible after the machine is in place.

Connections between the machine and the faucets are made with rubber hose. Figure 3-3 shows how the hoses should be attached. A rubber washer should be placed in the end of the hose at the machine connection. On the faucet end of each hose a rubber washer and screen should be used. The purpose of the screen is to filter small particles of debris out of the water and prevent them from clogging the inlet valves or other machine components.

Connecting the drainhose improperly is the most frequent mistake in washer installation. This hose attaches to the machine outlet with a Corbin clamp and connects to the sewer drain. Ideally, the drainhose should fit loosely into the drainpipe as shown in Fig. 3-4A. If the hose fits too tightly, as shown in Fig. 3-4B, it is possible that a syphon will form. The syphon action will cause the tub to continually drain and prevent it from filling with water. If it is not possible to allow a loose fit, then an antisyphon valve should be placed in the drain line as shown in Fig. 3-5.

Most manufacturers specify the drainpipe height at somewhere between 32 and 60 in. (Fig. 3-6). Below 32 in.

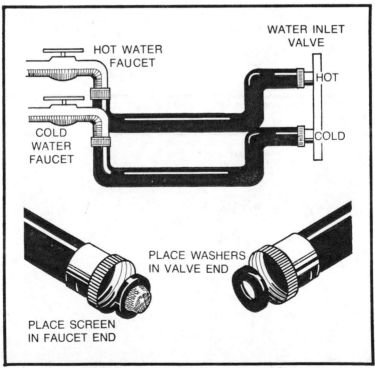

Fig. 3-3. Hose connections require washers and screens.

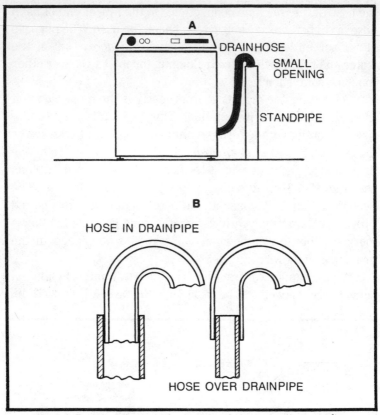

Fig. 3-4. Drainhose connections are made improperly most often.

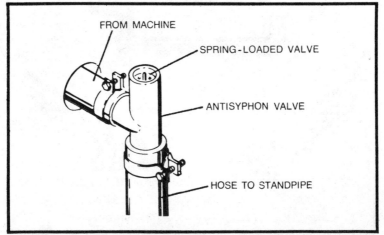

Fig. 3-5. Antisyphon valve prevents syphoning at drainhose connections.

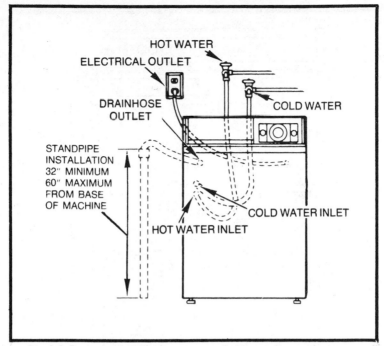

Fig. 3-6. Drainpipe height as recommended by most manufacturers.

gravity will start a syphon action. Above 60 in., the discharge pump will be overworked. The flexible drainhose should be looped and permanently fastened as shown in Fig. 3-7.

ELECTRICAL INSTALLATION

Unless you are licensed to make alterations to home wiring systems, you must require that the homeowner place an outlet of sufficient capacity in the immediate area of the machine. Most manufacturers recommend a separate 15A circuit for the washer. For the best results this circuit should be protected by a 15A breaker. All washers come equipped with a three-prong grounded plug on the power cord. This cord should be connected to a three-prong grounded outlet. The ground wire should never be altered. If a two-prong receptacle is encountered, most manufacturers recommend it be replaced with a three-prong service. When this is not possible (and local codes permit) the electrical hookup may be made with an adapter as illustrated in Fig. 3-8.

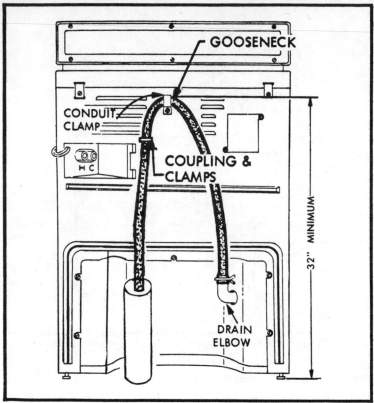

Fig. 3-7. Drainhose should be looped and fastened securely.

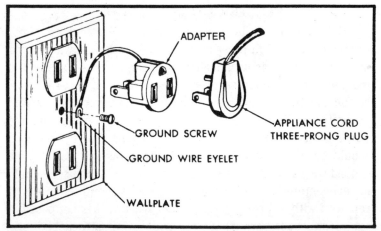

Fig. 3-8. Electrical hookups may be made with an adapter, but check the local code first.

SUDS-RETURN MACHINES

Some machines have a special feature to save hot water and detergent. These machines provide a method of discharging the suds into a tub at the end of the wash period and returning it to the machine to use with a second load. Figure 3-9 shows one manufacturer's recommended hookup for suds-return machines.

There are two discharge hoses on these machines. One is a conventional discharge hose, the other is the suds-return hose.

A tub of at least 20 gal. capacity is needed to store the suds during the rinse and spin-dry periods. Suds are discharged at the end of the wash period through the suds-return hose into a holding tub. During the suds-return cycle the sudsy water is drawn back into the machine through the suds-return hose.

After any washer installation, you should run the machine through a complete cycle of operation. Check hoses for leaks and the machine for proper fill, agitation, drain, and spin functions. Also check for vibration. This doublecheck procedure will reduce your callbacks.

LAUNDERING FACTS

Occasionally you will receive calls for washer service when the owner is unhappy about the general operation of the

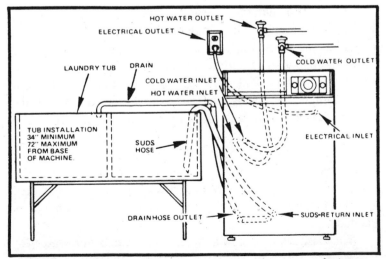

Fig. 3-9. Suds-return hookup as recommended by one manufacturer.

machine. In this instance the machine may go through all the separate operations; however, the customer is unhappy with the results: the clothes do not get clean or too much water is left in the machine. Although these complaints can sometimes be traced to a faulty machine, there are many times, especially if the machine is new, when the problem is incorrect use rather than incorrect operation. Such complaints must be handled with tact so as not to offend the customer. Explaining the machine's requirements and operation will instruct your customers in the proper use of their laundry equipment.

The Laundry Process

The laundry process is a complex operation consisting of both chemical and mechanical actions. Soiling is caused by particles of dirt caught in the fibers of the fabric. Some of these soiling agents will react to soaking, while others will not. Some will soften when heated, others will not. Some will react to chemicals and dissolve when saturated with certain detergents, others will not. The laundering process is a combination of methods designed to remove soil from most fabrics in the average home. Some fabrics will not respond to automatic home laundering, however, and require some other method of cleaning.

In a typical automatic washer the laundering starts by wetting the fabric. Most machines provide a means of adjusting the water temperature to the type of fabric being cleaned. Heavy soil in white or *colorfast* cotton fabrics will require hot water, while delicate fabrics and colors require a cooler temperature. Soils that are easily dissolved in water are loosened as the fabric is saturated with water. Once the clothing has been soaked in water at the proper temperature, chemical and mechanical action is applied to remove other soils. Detergents designed to loosen the bond between the fabric fibers and the soil are mixed with the water. Bleaches, water softeners, and fabric conditioners may be added as needed.

The wash water is now a blend of chemicals and water (of correct temperature) combined to loosen the soiling agents.

Some machines provide a *presoak* period that allows the clothing to sit in this solution for a short time, giving the chemicals an opportunity to work. After the clothing is thoroughly saturated, mechanical action is applied to remove other dirt. Older laundry methods required rubbing or beating the clothing to loosen the dirt. The automatic washer is more gentle and, in most cases, more thorough in this function. A paddle wheel or agitator in the tub oscillates to both move the clothing through the water and circulate the water over the clothing. This causes the clothes to bend and tumble in the water and rubs the fibers in the material against each other. Combined with water circulation in and around the material, this rubbing action loosens dirt caught between the fibers and suspends it in the washer.

The automatic washer allows a specific time for the wash cycle then forcefully expels the wash water and the dirt suspended in it. In most machines this is done by draining the water from the tub then spinning the tub at a high speed to sling the soiled water from the clothing. As this water is drained from the tub, particles of dirt and scum (formed by the chemical reaction of detergents, softeners, and minerals) suspended in the water cling to the external surfaces of the materials being laundered. To remove these the machine is filled once more with clean water. Again they are flexed in the water by mechanical agitation. After rinsing, the water is pumped from the machine and the tub is spun at a high speed to remove most of the water from the clothing.

The clothing should be *damp-dry* when the washing is complete. This is a hard term to define because it means different things to different people. Most machines are built to specifications that leave the clean, damp-dry clothing weighing approximately twice what it weighed before washing. Usually, a fairly accurate check of the machine's water-removing performance can be made by weighing the clothing before and after it is washed.

Water

Automatic washers require a large quantity of water to do their job properly. The exact amount, of course, depends on the amount of washing, the size of the machine, and the types of fabrics washed. An adequate supply of hot water, at 140–170°, must be maintained for good cleaning.

The hardness of the water will also affect the quality of the wash. Hard water has excessive amounts of mineral deposits that prevent the detergent from working. Also, the detergent reacting with these minerals leaves a scum in the machine that can clog water passages and leave a gray tint in the fabrics. Water hardness is usually expressed in grains per gallon (GPG). Between 0 and 3 GPG, the water is considered soft; 8 GPG is a hard-water average. Water with a mineral content of 19 GPG or more is excessively hard and requires some sort of treatment to counteract the effects of the minerals.

Chemicals

One of the oldest chemicals used in the laundry process is soap. Although this term is loosely used with all types of cleaning agents, soap is a compound of animal fats and oils. The animal fats combine with the dirt in the fabric and float to the top of the wash water. Soap works well in hot soft water. However, in cool or hard water the minerals and animal fats form a scum that floats on the wash water. When the water is drained from the machine some of this scum clings to the fabrics being laundered. This leaves the fabrics with a dingy appearance.

Detergents are *synthetic* soaps that overcome some of the problems associated with regular soaps. These cleaning agents are chemical compounds of several ingredients with specific purposes. Wetting agents in detergents cause the fabrics to "open up" and allow better saturation. Some of these same agents react with the dirt to keep it suspended in the wash water and prevent it from being redeposited on the clothes. Some detergents contain phosphates to aid in the wetting action and soften the water. Softeners react with the minerals in hard water to suspend them in the wash solution. Stabilizers (boosters) are used to control the sudsing. Various agents used to make fabrics appear brighter are found in many synthetic detergents. Some detergents contain enzymes that react with protein to loosen such stains as chocolate, milk, and blood.

90

If the water is too hard, much of the detergent is used to counteract the hardness. This reduces the cleaning ability of the detergent and causes a residue similar to soap scum. To overcome this the water can be treated with borax. Borax, as does many softeners, combines with grease and oil in the water to form a soap and softening compound. Some detergents contain borax or other chemicals as softeners. Water-softening chemicals that can be added to the wash water are also available. In extreme cases, however, the only remedy for hard water is a treatment system that will remove some of the minerals.

Bleaches used in home laundry products are either chlorine or oxygen compounds. Bleach will remove soils and stains and whiten fabrics that detergents alone cannot. These chemicals are more harsh than common laundry detergents and should be used with care. They should not be used on wool, silk, or certain blended and noncolorfast fabrics.

In addition to water softeners, there are also fabric softeners on the market. These products make the fabrics softer, fluffier, and easier to handle. They also reduce the tendency of synthetic fabrics to cling due to static electricity.

With today's synthetic fabrics, starch is not used as frequently as it was in the past. Starch is made from a variety of ingredients (cornstarch, wax, and resins, for example) and is used to give a smooth appearance to fabrics. Similarly, *bluing*, once used to brighten the fabrics, is being replaced by fluorescent dye-brighteners in detergents

Problems

As mentioned previously, some laundry problems can be eliminated by proper use of the machine and the chemicals used in the laundry process. One common problem is a dull or gray appearance in things that should be white. This is usually caused by dirt or oils being redeposited on the clothing when the water is drained from the tub. Improper sorting, overloading, insufficient water, insufficient detergent, or hard water can cause gray fabrics.

Yellowing can result from using too much bleach on certain fabrics, such as wool or silk. If this is the case the material will usually retain a strong chlorine odor after washing. Manganese in the water system can also cause bleach to yellow fabrics. A filter is usually the only solution to this problem.

Many automatic washers contain filters to remove lint. The effectiveness of these filters is reduced if the machine is overloaded, clothing is not properly sorted, the machine is improperly operated (cycle too rough for fabric being washed), the water level is too low, or fabric softeners are used improperly. Fabric softeners should be used only in the final rinse.

Overloading will frequently damage the materials being washed. Washers have a definite safe working load that should not be exceeded. Most machines are now designed for 14-, 16-, and 18-pound capacities.

SUMMARY

Many laundry problems can be overcome by proper installation and proper use of laundry equipment and chemicals. Each new machine comes with complete instructions. Always refer to the manufacturer's instructions and those contained on the packages of laundry chemicals. If you suspect that improper use is the cause of a customer's complaint, tactfully ask the customer to demonstrate the problem with an actual wash load. Observe how the clothes are sorted, the machine is loaded, etc. If mistakes are observed use the manufacturer's literature to suggest corrections.

Chapter 4

Water Systems

To aid in understanding the operation of automatic washers the entire machine can be divided into three major systems: mechanical, electrical, and water. The water system with its associated components in a modern automatic washer is a complex pumping and control mechanism. Understanding the flow and routing of water within the machine will take you a long way in learning to diagnose troubles in machine operation.

The main purpose of the water system is to fill the machine to the correct level with water of the desired temperature and to remove that water from the machine at the proper time. In addition, depending on the features designed into the machine, the water system may also perform such functions as lint removal, suds storage, and bleach or detergent dispensing. These added features will vary from model to model even within the same brand. Therefore, if you are not familiar with the operating characteristics of a particular machine, consult the manufacturer's service literature. In many cases the owner's operating instructions are also helpful.

COMPONENTS

Many types of valves, pumps, filters, and other water-handling devices can be found in the water system of

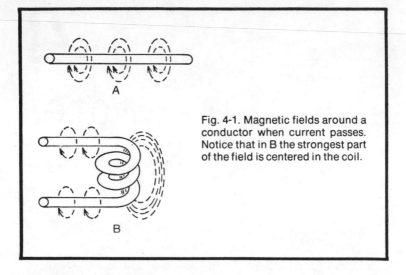

Fig. 4-1. Magnetic fields around a conductor when current passes. Notice that in B the strongest part of the field is centered in the coil.

any automatic washer. Some of these devices are electrically activated while others are mechanically controlled. Defects in these components can result in a variety of troubles. Understanding these functions and operations will help in troubleshooting automatic washers.

Solenoids

Solenoids do most of the work of controlling an automatic washer. These devices are nothing more than magnets that are electrically controlled—electromagnets. Such magnets are made by winding several turns of wire around a core that holds the windings in place. When current flows in a conductor a magnetic field is established about that conductor as shown in Fig. 4-1A. Winding this conductor in a coil as shown in Fig. 4-1B concentrates the magnetic field in one central location. The strongest part of the magnetic field will be centered in the coil.

This magnetic field can be made to exert a pulling force on magnetic metals such as iron. Figure 4-2 is a simplified diagram of a solenoid. An iron rod inserted in the coil of a solenoid is held down by a spring attached to the frame of the machine in which the solenoid is used. The switch (S1) shown in the *open* position connects the coil to the 120V power supply. In this condition the spring keeps the iron rod pulled downward. When the switch is closed the magnetic field generated in the coil exerts a pulling force on the iron rod, striking the spring. If the switch were opened again the magnetic field would collapse and the spring would pull the rod back to its *rest* position.

The action of the solenoid can be used to control various mechanical components of an automatic washer. An electric clock is used to apply power to the solenoid coil at the proper time. The solenoid can vary in size and pulling strength from a very small coil with only a few inch-ounces of pull to a very large coil exerting several foot-pounds.

Inlet Valves

One use of the solenoid is operating the inlet valve that supplies water to the machine. The valve, shown in Fig. 4-3, can aid in understanding the operation of these devices. A flexible diaphragm across the valve port acts as a seal to prevent water flow when the solenoid is deenergized. The spring forces the plunger downward to hold the solenoid against the valve

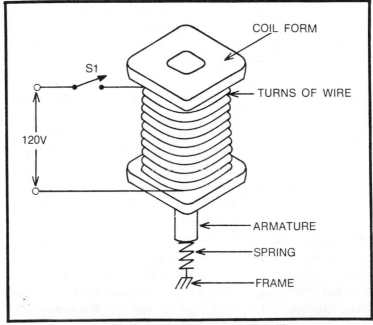

Fig. 4-2. Simplified diagram of a solenoid.

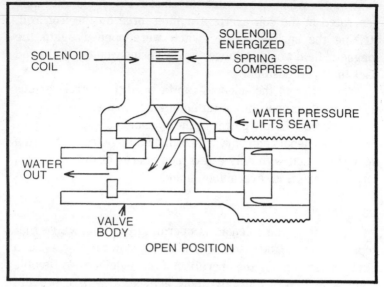

Fig. 4-3. A water inlet valve in the open position.

seat. Water enters the valve through the inlet port and is
filtered by a small screen that prevents debris from entering
the valve. Water, at line pressure, exerts an upward force on
the diaphragm that tends to lift the diaphragm from its seat.
The spring pressure pushing downward on the plunger holds
the diaphragm in place. A small hole in the diaphragm allows
water to enter the chamber above the diaphragm. Since water
pressure is exerted on both sides of the diaphragm the amount
of spring pressure required to keep the valve closed is
reduced. The outlet port of the valve connects to tubing and
hoses that route the water to the tub of the machine.

Figure 4-3 shows what happens when the solenoid is
energized. The magnetic pull of the coil lifts the plunger off its
seat. Water pressure then forces the diaphragm off its seat and
water flows from the inlet port. This pushes the diaphragm
upward. The water can now flow over the valve seat and out
through the flow washer. The flow washer is a special rubber
washer with a hole in its center. One side of this hole is
rounded (the side facing against the flow of water). This
washer is constructed to match the valve to the machine by
controlling the rate of waterflow.

The diaphragm lifts from the valve seat and water trapped in the opening above the diaphragm drains through the pilot hole in the center of the diaphragm. This water then flows with the incoming water out through the flow washer and into the inlet tubing and hoses.

Power is removed from the solenoid coil to cut off the flow of water. Spring pressure forces the plunger down, striking the diaphram and pushing it against the valve seat. Water entering the bleeder hole equalizes water pressure on both sides of the diaphragm, and the spring holds the valve closed.

Typically, these are dual valves having two inlets (hot and cold), two solenoids, and one outlet. Such a valve is shown in Fig. 4-5. Sometimes called a mixing valve, it can control the flow of water to furnish hot, cold, or warm water. Two hoses connect the valve to the hot and cold water faucets located behind the machine. Notice that both inlet valves feed into a common mixing chamber and through a single opening into the machine. This two-coil valve, shown in the *closed* position, has water pressure from each inlet port present at the diaphragms. In this instance no power is applied to either coil, both valves are closed, and no water flows into the machine.

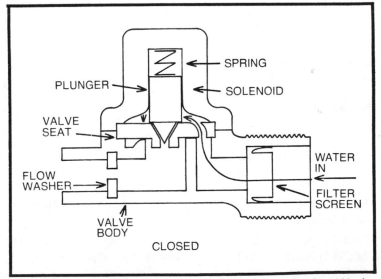

Fig. 4-4. A water inlet valve in the closed position when the solenoid is de-energized.

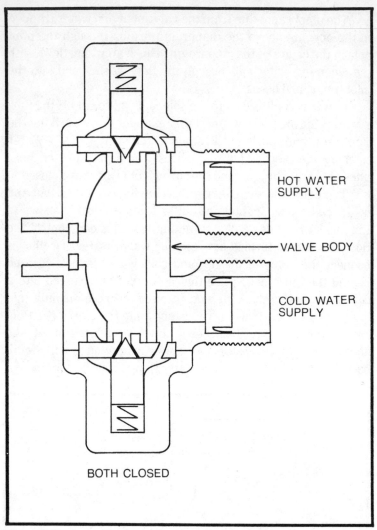

HOT WATER
SUPPLY

VALVE BODY

COLD WATER
SUPPLY

BOTH CLOSED

Fig. 4-5. The two-coil valve in the closed position. This valve controls the flow of hot, cold, or warm water.

In Fig. 4-6 power has been applied to the cold water coil but not the hot water side. The cold water coil lifts the associated plunger, and water pressure forces the diaphragm from the valve seat. Water can now flow from the cold water faucet through the cold water side of the valve and into the mixing chamber. Hot water flow, shown in Fig. 4-7, is similar to that described for the cold water side. With power applied only to the hot water coil, the cold water diaphragm remains seated and only hot water enters the machine.

To fill the tub with warm water, both coils are energized and both diaphragms lift. This allows both valves to open and both hot and cold water enter the tub as shown in Fig. 4-8. With both diaphragms off their seats, hot and cold water enter the mixing chamber. Although this water is called warm (hot and cold mixed) the temperature depends on the actual temperature and pressure of the hot and cold water supplies. The warm temperature can be adjusted by the customer at the

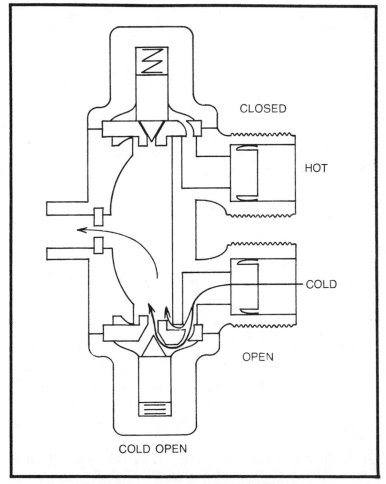

Fig. 4-6. A two-coil valve admitting cold water.

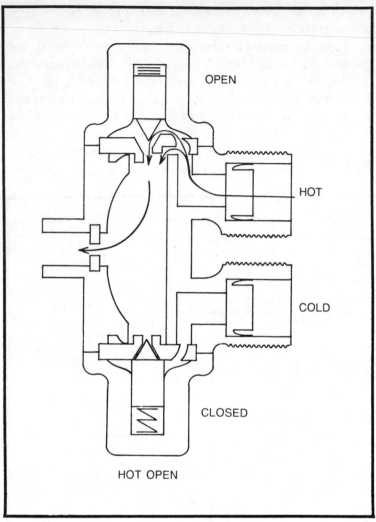

OPEN

HOT

COLD

CLOSED

HOT OPEN

Fig. 4-7. A two-coil valve admitting hot water.

cold water faucet. Usually, cold water pressure is somewhat higher than hot water pressure. Therefore, best results are normally obtained by opening the hot water valve all the way and turning the cold water valve enough to achieve the desired temperature.

The actual appearance of the water inlet valve will vary from make to make and model to model. Usually, they operate similar to the description given here. A variation of this is the

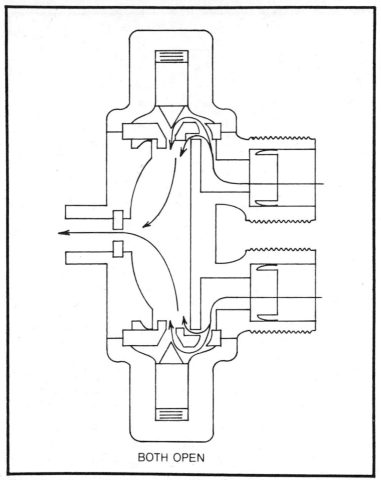

BOTH OPEN

Fig. 4-8. A two-coil valve admitting warm water. Both coils are now energized.

three-coil valve shown in Fig. 4-9. In this instance a separate coil is energized for warm water. Notice that both hot and cold water is supplied to the warm-water diaphragm. Small check valves may be built into these ports to prevent the hot and cold supplies from mixing except when warm water is selected. A check valve is simply a device that limits water flow to one direction. Check valve operation is explained later in this chapter.

Some models control water temperature to a finer degree than that provided by the valves just explained. In such cases

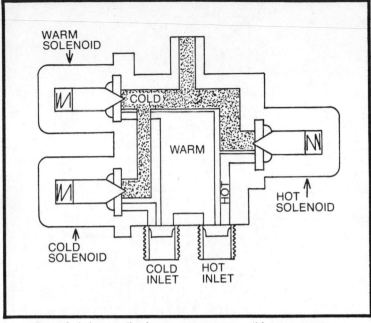

Fig. 4-9. A three-coil valve uses a separate coil for warm water.

thermostatically controlled valves are used. These valves are made with a thermal element located in the mixing chamber. When warm water is selected, water flowing over the thermal element causes it to expand and contract. This regulates the mixing of hot and cold water to achieve the desired temperature. A control permits adjustment of the warm water setting through several stages. Some machines are equipped with a three-coil variation of the thermostatically controlled inlet valve. Regardless, the function is the same.

Service to inlet valves is usually limited to replacement. Although kits that furnish the internal parts needed to repair these valves are available, experience has proven that in most cases it is better to replace the valve rather than repair it. One common trouble developing in inlet valves, especially when the machine is used in an area where the water is very hard, is leaks. Small mineral deposits develop on the valve seats that prevent it from seating properly. This allows water to flow into the machine at all times, even with power disconnected. If water enters when the power cord is unplugged, the inlet valve should be replaced. These valves may also stick in the closed position.

The valve can be tested with an ohmmeter. Measure the resistance across each coil. Although the exact value of this resistance will depend on the model, it typically runs around 500Ω. A test cord can also be used, but cautiously since hot leads will be exposed.

One method of bench testing is shown in Fig. 4-10. Using a Y-hose equipped with three female fittings, the valve can be

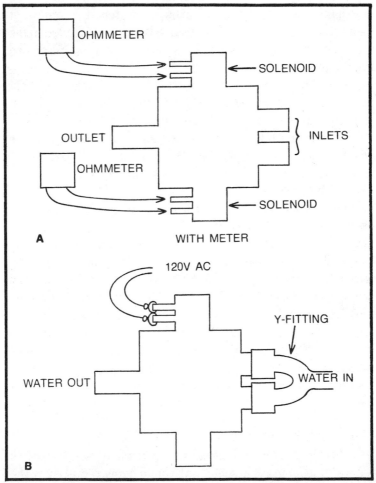

Fig. 4-10. Testing water inlet valves on the bench.

connected to a cold water supply as shown in Fig. 4-10. The test leads are two meter probes connected to a conventional extension-cord plug. When this cord is plugged into a wall receptacle 120V is present, so be careful to avoid touching these probes. With water supplied to the Y-fitting touch the probes to the terminals of one of the coils. That port should open and allow water to flow. This test should be done at a workbench with a sink to catch spilled water. Remove the test probes if water flows. The water should stop. Use the same procedure to check the other coil. This test will prove beyond any doubt whether the valve is functioning properly. However, due to the presence of hot terminals and water, it is somewhat dangerous. A safer method is to leave the valve installed in the machine and use the troubleshooting procedures described later in this book.

The ohmmeter test can be used to check the electrical properties of a valve during bench testing. This is a safe, fast way to locate open coils. A simple test that reveals leaky valves is to blow hard into the outlet port of the valve and note whether there is any leakage at either inlet.

Dispensers

Some machines use solenoid valves to control the mixing of detergents, bleaches, or other additives. Typically, these dispensers have a container in which the additive is placed. A solenoid valve between the container and the hose supplying the liquid to the tub prevents the agent from reaching the tub before the correct time. The liquid will gravity-drain into the tub when the coil is energized.

In some cases the solenoid valve routes inlet water to a dispenser. The water then carries the rinsing agent (or other product) into the tub. Regardless, if solenoid valves are used, operation and testing is similar to that of the water inlet valves.

Float and Pressure Switches

Float and pressure-operated switches are used to control the water level in the machine. Although being replaced by the pressure switch on most machines, simple float switches, such as that shown in Fig. 4-11, are sometimes used. The electrical contacts of the switch make and break the circuit to the water-filling components and, in a sense, tell the machine the present water level in the tub.

The float system uses a water column, either the actual water level in the tub or water in an open column as shown in Fig. 4-11, to trip a microswitch. The switch may be normally closed (as shown) or normally open, depending on the manufacturer's design. Normally-closed contacts complete the electrical circuit to the inlet valves when the tub is empty. The water level in the column rises as water enters the tub. When the water reaches the desired level the float causes the switch contacts to open, in this example removing power from the water inlet valves. In other instances where the contacts are normally open, the float causes the contacts to close. This moves the timer from the *fill* position and removes power from the inlet valves. In either case the result is the same—the water is shut off.

A more often used water-level control system is shown in Fig. 4-12. A column of air is used instead of water. Air caught in the tube is compressed as water pushes upward in the tube. Since there is no place for the air to escape, the higher the

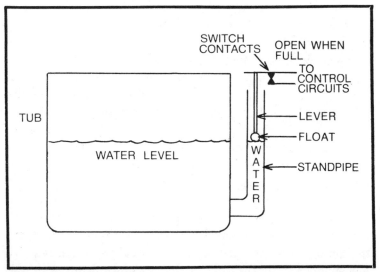

Fig. 4-11. A typical float switch used to control the water level in an automatic washer.

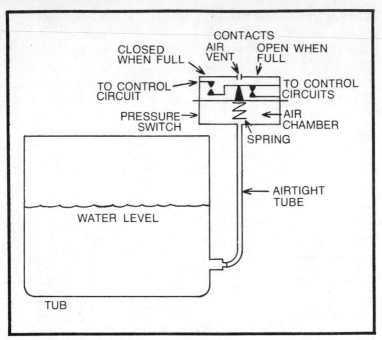

Fig. 4-12. A typical pressure switch is used more often in modern machines.

water level in the tub the more pressure is exerted on the switch.

Figure 4-12 will aid in understanding how the switch functions. An airtight diaphragm held in place by a spring places a specified amount of pressure on the air in the tube. The water rises causing air pressure to increase, and the force against the diaphragm increases. When the water reaches the proper level the diaphragm is forced against the movable contact of the switch. In this particular example a two-pole switch is used. This is a common arrangement. The hot side of the 120V supply is connected to the movable contact. When the tub is empty the movable contact is made with the lower stationary contact. At the instant air pressure overcomes the diaphragm-spring pressure, the movable contact breaks with the lower contact and makes with the upper contact. Usually, this action is used to remove power from the water inlet valve and apply power to the timer motor. As water drains from the tub, spring pressure returns the switch to its original position.

There are three principal types of pressure-operated water level controls. The one just described is the single-level type; that is, there is only one pressure corresponding to a specific water level that trips the switch. A second type has a selector switch with two or more discrete positions that mechanically control spring pressure applied to the diaphragm. This allows the operator to select the desired water level to match the wash load. The third type of control also uses varied spring tension to select water level; however, this switch has an infinite number of selectable levels. The control lever can be set at any water level, rather than only two or three preselected spots, between *low* and *high*.

Don't confuse these operator controls with the calibration control (inaccessible to the user) located on the switch. The calibration control is used at the factory to set the point at which the diaphragm operates the switch. Most manufacturers do not recommend adjusting the level calibration in the field. Instead, if the switch is out of calibration it should be replaced. Repairs or adjustments to these switches are usually short-lived.

Testing these switches can best be done with the switch installed in the machine. After disconnecting the machine from its power source make sure no water is left in the tub. Using the manufacturer's wiring diagram or service data, locate the pressure-switch contacts that should be closed during the filling of the machine. Put the ohmmeter on the $R \times 1$ scale and measure the resistance between the two contacts as shown in Fig. 4-13. It should be zero. Now measure the resistance between the contacts that are open during filling. There should be infinite resistance. Fill the machine to the prescribed level and recheck the resistances. They should be reversed. If not, replace the switch.

Check Valves

Figure 4-14 is a cutaway view of a simple check valve. This valve can be used to allow the flow of liquid in one direction, while preventing it in the opposite direction. The check spring holds the hinged gate against the seat when no pressure is applied. When water under pressure is connected to the inlet,

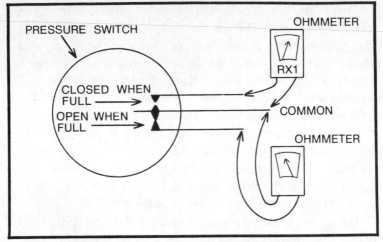

Fig. 4-13. Testing the pressure switch with an ohmmeter.

the gate raises against the spring pressure and allows the water to flow through the valve. If water pressure is applied to the outlet the pressure forces the gate against the seat and prevents reverse flow.

Check valves may be tested after removal from the machine by blowing into the outlet port and checking for any leak at the inlet port.

Filters

Water filters used to remove lint from the wash and rinse waters vary from make to make. Some filters, attached to the

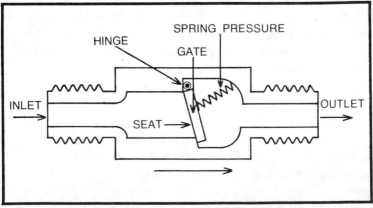

Fig. 4-14. A cutaway view of a typical check valve.

agitator post. are basket arrangements through which the water is recirculated. The filter basket that catches the lint is a fine mesh screen through which the water passes. The operator removes the basket after the washing is complete and removes the lint.

A more automated version of the recirculating filter is demonstrated with the aid of Fig. 4-15. The check valve near the top of the tub allows water to be drawn from the top of the tub (where the lint is floating) to be recirculated into the bottom of the tub. Lint in the water is trapped by the screen and held in the filter until the water is discharged. During the discharge cycle the pump runs in the opposite direction. Water is now drawn from the bottom of the tub and flows in this reverse direction through the filter. This flushes the trapped lint out of the filter. As the water reaches the check valve the pressure seats the valve in the tub opening and opens the discharge port to route the water out of the machine. In some cases solenoid-operated valves are used to reroute the flow of water.

Pumps

Water enters the machine under water-main pressure. Once the water inlet valve turns off the water, however, the

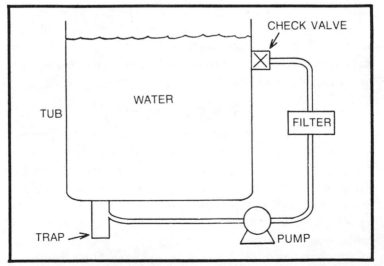

Fig. 4-15. A recirculating water filter is used to trap lint.

machine needs some method of internally moving the water. A pump is used to recirculate the water in some models during wash and rinse cycles. All machines use a pump to remove the soiled water from the machine. Although these pumps differ in appearance they operate on the same principle.

On some older machines a mechanical device is used to engage the pump pulley with the drive belt during the pumpout phase of the wash cycle. Most modern machines use a pump driven constantly in all modes when the motor is running. During the agitation cycle these pumps are sometimes driven in reverse to prevent pumping. During pumpout and spin cycles the motor reverses and pumping takes place. On other makes a series of check valves and solenoid-control valves determine whether any actual pumping takes place.

Although pump repair kits are available for some models, it is generally a better practice to replace the pump as a unit whenever trouble occurs. The difference in cost between the repair kit and a rebuilt or new pump will usually be less than the labor charge associated with installing a kit. Pump operation can best be checked during actual machine operation.

Some machines are equipped with a so-called two-stage pump. Actually, this is two separate pumps driven by a common rotor. When the pump is turning clockwise, one impeller is turning in the direction to pump water while the other impeller is turning counterclockwise and has no effect on water flow. When the rotation is reversed the opposite impeller begins to move the water while the other turns backward.

TYPICAL WATER SYSTEMS

Although the water systems employed in most automatic washers have many similarities and use many components that perform the same function, there are significant differences from model to model. To use only one or two types for explanation purposes here would shortchange the reader by not providing a broad enough knowledge of water-system operations. Conversely, to cover every type in detail would be beyond the scope of any single volume. What we have attempted here is to give an overall picture of several types of water systems found in today's machines. You should be able to understand the water flow system in any automatic washer by studying either the system in an actual machine or the manufacturer's service literature.

Nonrecirculating

The best system to study first is the nonrecirculating, nonfiltering water system. This is the simplest of all the systems and, since it makes no provision for filtering the wash and rinse waters, it is usually found only on price-leader models. Such a basic system is shown in Fig. 4-16.

Five major components and their associated hoses, clamps, and seals are used in this system. The water inlet valve controls the flow of hot and cold water to the tub. A single hose connects the inlet valve to a small flume at the tub. As selected by the electrical system, hot and cold water is

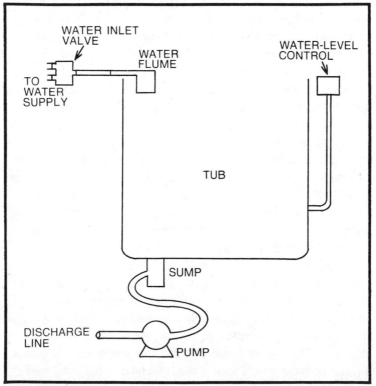

Fig. 4-16. A typical basic water system found in economy model machines.

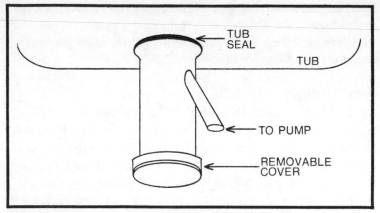

Fig. 4-17. The sump or trap is used to catch foreign objects before they enter the pump.

admitted from the house plumbing and routed through the hose to the flume where it pours into the tub. The tub, sealed at all openings from the mechanical system, holds the water during wash and rinse cycles. The fill control senses the water depth in the tub and, by electrical signal, tells the timing controls when the tub has reached the selected level.

At the time designated by the electrical control system, the pump will begin to remove the water, taking suction at an opening in the bottom of the tub. In order to protect the pump from foreign objects a sump device is installed at the discharge opening. The sump (manifold, or trap, as it is sometimes called) is constructed so that should a heavy object (a small stone, for example) enter the discharge outlet, the weight of the object will cause it to sink to the bottom of the sump. As shown in Fig. 4-17, the pump takes suction above the sump at a distance sufficient to prevent objects from being drawn into the pump. Some of these pumps are built with a removable plug that allows debris to be removed.

A machine using this type of water system would go through an operating cycle similar to that shown in the timing diagram of Fig. 4-18. The complete washing cycle is comprised of three major time periods: wash, rinse, and spin. These periods include one or more discrete operations such as pump, fill, etc. In some cases each period can be divided into smaller time slots that designate specific functions, such as agitate, fill, pumpout, etc. Using the diagram, a typical cycle is explained as follows.

After the clothes are placed in the machine and the detergent is added, the operator starts the machine. At this time the water-level control senses an empty tub and signals the water inlet valve. Opening the inlet valve starts the fill operation. During this period the only thing happening is water pouring into the tub. The main drive motor is deenergized and the pump is not running.

Upon completion of the filling operation the water-level switch is moved to its full position by pressure in the line from the tub. The water inlet valve is closed and the electrical system signals the mechanical system to start the agitation cycle. The timer is usually started at this point and the machine will wash (agitate) the clothes for a time determined by the timer control.

At the completion of the agitation cycle the electrical controls stop the agitator and start pumping water out of the machine. In this example this is accomplished by reversing the main drive motor rotation which causes the pump to turn in the proper direction and expel water. When the water level decreases to a point where pressure to the water-level control

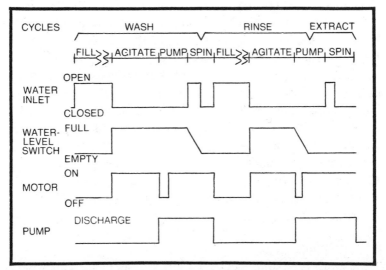

Fig. 4-18. A water system timing diagram showing a typical operating cycle.

is decreased, the control will shift to the spin cycle. As the spin basket revolves at a high speed, water is slung from the clothes and, since the pump is still operating, pumped from the machine. The length of time that the basket spins is determined by the electrical timer. In most cases the timer will intermittently energize the water inlet valve during this spin cycle. This sprays the spinning clothes with water and removes soap scum and lint. (For simplicity, this intermittent opening and closing of the water inlet valve has been omitted from the timing chart.)

Upon completion of the first spin cycle the driver motor is stopped and the water-level control (now on empty) opens the water inlet valves to fill the machine with rinse water. As during the first fill period, the motor and pump remain inactive during this time. Later we will see that in most machines the timer also does not advance during the fill operation. Most machines have provisions for selecting the rinse water temperature as well as wash water temperature. This is a function of the electrical control circuits and will be explained in a later chapter.

Once the water has reached the correct level, air pressure built up in the water-level control tube switches the control to the *full* position. The timer shifts the solenoids controlling the machines mechanical components to the agitate position. Again the clothes are agitated for a specified period of time. Remaining soil, wash water, and suds residue is removed from the clothes and left suspended in the rinse water. After this cycle the motor is off for a short time.

The timer now starts the motor in the pumpout direction and the pump begins to remove water from the machine. When the water level drops low enough, the water-level control shifts to the empty position, allowing the controls to spin the tub. The pump continues to remove water expelled from the clothes. When the cycle is complete this spinning action continues for a length of time controlled by the timer.

Recirculating

The water system illustrated in Fig. 4-19 has a somewhat different pump, but follows the same general operating cycle as that discussed in conjunction with Fig. 4-16. This particular pump is designed to pump water regardless of which direction the motor turns. The pump is equipped with a valve that may be controlled by a solenoid or the direction of pump rotation. When solenoid controlled, the timer signals the solenoid as to whether the machine is in a wash, rinse, or pumpout operation. The solenoid is energized during wash and rinse operations, and positions the valve to block the discharge line and open the recirculation line.

Water drawn from the tub during wash and rinse cycles is returned to the tub where it exits through a lint filter and reenters. This filter is simply a strainer that catches the lint suspended in the water. The operator should clean the filter after each wash. In the pumpout periods the solenoid is

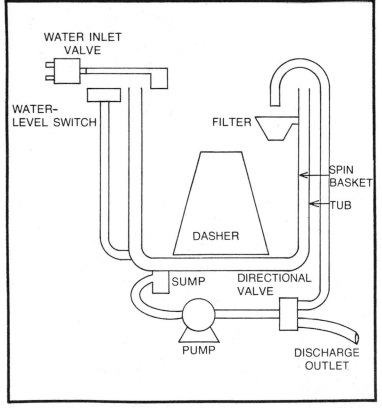

Fig. 4-19. A recirculating water system pumps water regardless of the rotation of the motor.

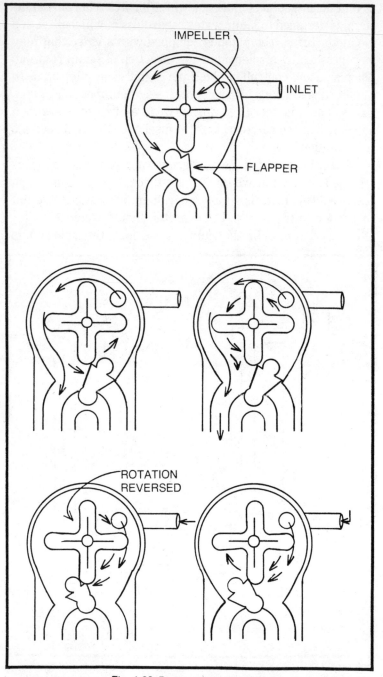

Fig. 4-20. Pump-valve operation.

deenergized. A spring pulls the valve lever in a direction that closes the line to the filter and opens the discharge line. Now when the pump rotates, water is expelled from the machine.

Some machines, operating in the same manner, use the direction of the pump of Fig. 4-20. Assume the valve is in position to discharge water from the machine. At the end of the fill operation the motor starts and the pump begins to recirculate the water. The pump starts to rotate and the water, under pressure from the impeller, strikes the valve flipper. This moves the flipper off its set where the impeller can actually engage the flipper and position it to block the discharge hose. The flipper is then held in place by spring tension and water pressure.

The pump reverses rotation to discharge water. Again water pressure moves the flipper off its seat far enough for the impeller to strike it and position it across the recirculation line. Now water is drawn from the tub and returned through the filter.

Dasher Filter

Another filter uses a nonrecirculating water system similar to that in Fig. 4-16. Operation of this system is similar except for the filtering action that takes place during wash and rinse cycles.

When the agitator in Fig. 4-21 is dashed back and forth in the water, the wash (or rinse) water is forced up through the agitator by valves inside the agitator. Water spills out through the upper openings and is strained through the filter encircling the agitator. As with the previous filter, the operator cleans the filter basket after each wash.

Self-Cleaning Filter

Some manufacturers have introduced machines with a self-cleaning filter system. These machines eliminate the necessity of cleaning the lint filter. Figure 4-22 is an illustration of such a system.

This system uses a check valve in the side of the tub just below the normal water level. When pressure is applied to the filter side of the valve, it closes the tub opening and opens the

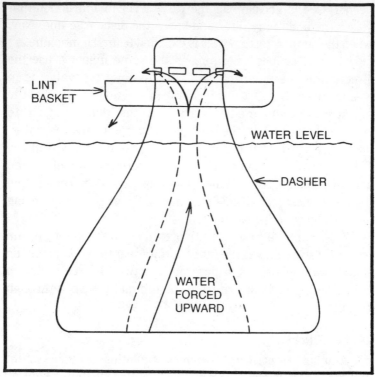

LINT
BASKET

WATER LEVEL

DASHER

WATER
FORCED
UPWARD

Fig. 4-21. A dasher filter operates during wash and rinse cycles.

discharge line. Suction applied to the filter side of the line draws water from the tub. Since most lint floats near the surface of the water, these systems are usually more effective than those drawing recirculating water from the bottom of the tub.

The cycle of operation is basically the same as previous systems except for filter operation. When the pump is rotating in the recirculate direction, water is drawn from the tub through the check valve, the filter and the pump, then pushed back into the bottom of the tub. Lint suspended in the water is trapped on the check-valve side of the filter and held until the water is discharged from the machine. The motor reverses, and the pump is driven in the opposite direction and water is drawn from the bottom of the tub. Water is forced through the filter in the opposite direction. This carries the lint from the filter to the check valve. Water pressure causes the check valve to seal the tub opening, and the lint-laden water is pumped from the machine.

One variation of the system shown in Fig. 4-22 uses a drive motor turning continuously in one direction. The direction of water flow is controlled by a flapper valve in a small manifold attached to the pump. During wash and rinse cycles, the flapper valve routes the water from the check valve through the filter and pump and back into the tub through the trap. This recirculates the water and lint in the filter. During pumpout, the flapper valve reroutes the water in the pump manifold, taking suction at the trap and pumping back through the filter and out through the check valve.

Suds-Return Systems

Some manufacturers offer a special feature on their machines that allows the sudsy water to be stored at the end of

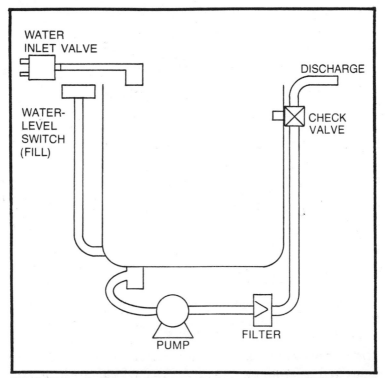

Fig. 4-22. A self-cleaning filter system eliminates the need to clean the filter after each wash.

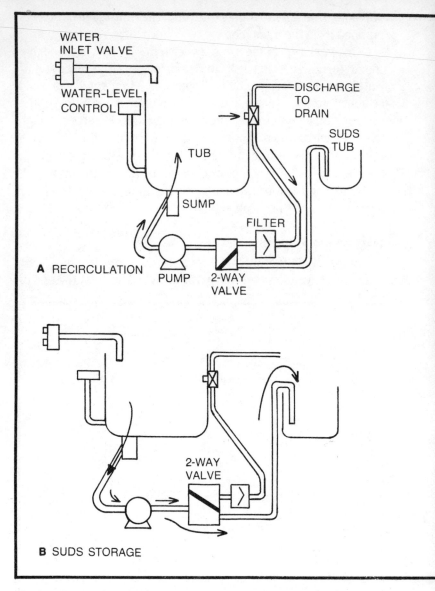

A RECIRCULATION

WATER INLET VALVE

WATER-LEVEL CONTROL

TUB

SUMP

DISCHARGE TO DRAIN

SUDS TUB

FILTER

PUMP

2-WAY VALVE

B SUDS STORAGE

2-WAY VALVE

the wash cycle and returned to the machine during the next wash load. In addition to special features built into the machine, a suitable tub arrangement must be provided for temporary storage of the sudsy water. Figure 4-23 is a simplified diagram of a complete water system with these benefits.

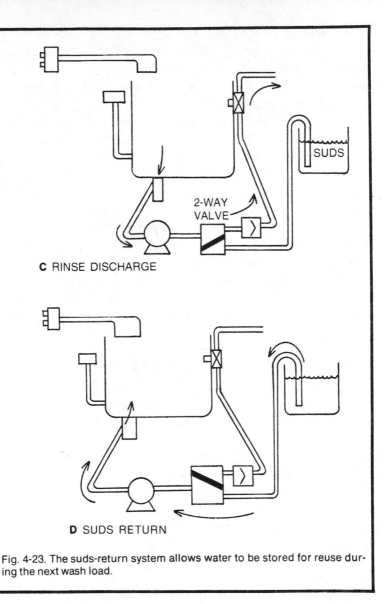

C RINSE DISCHARGE

D SUDS RETURN

Fig. 4-23. The suds-return system allows water to be stored for reuse during the next wash load.

Initial filling is accomplished using the same methods as other systems. A special two-way valve is inserted in the line between the pump and self-cleaning filter. During the wash cycle, water is recirculated by the pump which draws water through the check valve, filter, and the upper part of the two-way valve, as shown in Fig. 4-23A. At the end of this wash phase, solenoid action closes the upper part and opens the lower part of the two-way valve mechanism. When the pump attempts to expel water from the machine in the normal manner, the valve reroutes the water to a hose that sends sudsy water to an external storage tub (Fig. 4-23B).

The electrical timing system repositions the valve for the rinse cycle. Now the system acts like a conventional recirculating filter system (Fig. 4-23A). Upon completion of the rinse cycle the two-way valve is held and discharge pumping starts. With the upper port open and the lower port closed on the two-way valve (Fig. 4-23C), water is expelled through the filter and check valve in the usual manner.

When the next load of clothes has been placed in the machine and the machine starts to fill, instead of filling through the fill valve in the normal fashion, the pump is started and the lower port of the two-way valve opens (Fig. 4-23D). Sudsy water is then drawn from the tub and returned to the machine.

LAUNDRY AGENT DISPENSERS

Several different methods of automatically dispensing laundry additives are used on modern automatic washers. Solenoid valves, mentioned earlier in this chapter, are used by several manufacturers to add detergent or bleach at just the right time. These reservoir-type dispensers consist of tanks emptied (at the proper time) through solenoid valves into the wash water. The tanks are filled by the operator according to manufacturer's instructions. At least two other popular methods are used to achieve the same results. These two systems are discussed here.

Water Dispenser

Turn to Fig. 4-22 and review the water flow in that recirculating system. Figure 4-24 is a modification of that system that provides a means of automatically injecting additives to the wash water. The major difference between the two systems is the use of a two-stage, single-direction pump with a flapper valve in the upper stage. Actually, the pump is two separate pumps attached to a single shaft. The pump is constantly turning in one direction. The upper stage of the pump functions identically to the flapper-valve pump discussed in conjunction with Fig. 4-22. During wash and rinse cycles the upper stage draws suction at the spring-loaded check valve in the side of the tub, pulls water through the filter, and sends it back to the tub through the trap. At pumpout times the flapper valve on the upper stage causes the pump to take suction at the trap and expel water through the check valve. The check valve seals the opening in the side of the tub and allows water to leave through the discharge hose.

The lower-stage pump delivers a smaller volume of water than the upper-stage pump. During the wash cycle, at the time detergent should be added, the solenoid-operated check valve opens by an electrical signal from the timer. The upper-stage pump moves water from the spring-loaded check valve at the

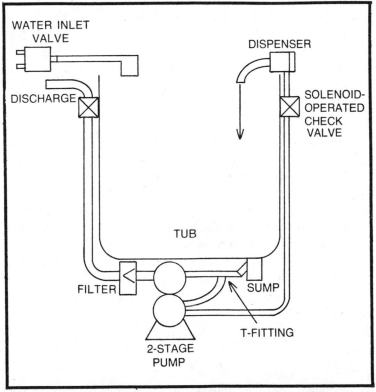

Fig. 4-24. A water-circulation detergent dispenser injects additives into the wash water.

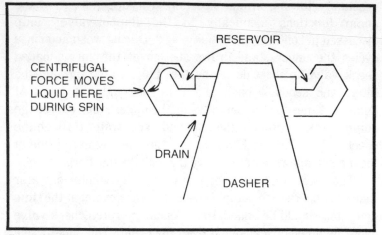

RESERVOIR

CENTRIFUGAL
FORCE MOVES
LIQUID HERE
DURING SPIN

DRAIN

DASHER

Fig. 4-25. This rinse-conditioner dispenser operates on the centrifugal force generated by the spinning basket.

side of the tub and supplies pressure at the trap. The lower pump takes part of this water and routes it through the now-open solenoid check valve to the detergent dispenser. The flow of water through the dispenser carries the detergent to the tub.

Centrifugal-Force Dispensers

One manufacturer makes use of the centrifugal force generated by the spinning basket to inject rinse conditioners into the tub. A donut-shaped dispenser is fitted to the agitator post as shown in Fig. 4-25. The operator loads the reservoir prior to the start of the wash cycle. The rinsing agent is held in the reservoir during the wash cycle. During the wash/spin cycle the centrifugal force of the spinning basket draws the rinse conditioner over the wall of the reservoir and holds it to the side of the dispenser. As the spinning basket coasts to a stop at the end of the wash/spin cycle, the centrifugal force decreases and the conditioner flows out the bottom of the dispenser.

SERVICING

A faulty water system in an automatic washer is not always readily identified, except when there is leakage. At first you may suspect the electrical or mechanical components as the source of a trouble, but later investigation might prove you wrong. Some technical publications advise you, as your first step in troubleshooting, to isolate the trouble to one of the three major systems. This may be difficult, if not impossible, at times. Suppose the solenoid check valve in Fig. 4-24 is allowing water to be pumped through the detergent dispenser during the spin cycles. Where is the fault? Is it the water system? Is it mechanical? Electrical? The solenoid check valve can be said to be part of all three systems. It is electrically operated (the timer or solenoid coil could be faulty). It is also a device that converts electrical energy to mechanical motion (the linkage could be frozen). And it controls the flow of water (the valve seat could be leaking).

Here we recommend a functional approach to troubleshooting. Treat the entire washer as one system. Break it down into operations such as fill, wash, spin, etc. Observe the cycle of operation and determine what function is not properly completed. Limit your checks to components that affect that function. (This method of troubleshooting is outlined in greater detail in a later chapter of this book—after chapters on the mechanical and electrical systems.)

One of the best troubleshooting aids is a thorough understanding of a machine's operation. When studying any machine pay particular attention to the operating cycle. Knowing what it should do and when it should do it is the key to troubleshooting.

Repairs to water-system components are usually limited, and parts replacement is the most accepted form of repair. Pumps, valves, and water-level switches can be repaired on some models but the cost of labor often justifies replacement with a rebuilt or new component. In our shop, we use a 50% rule. Whenever the labor cost runs to 50% of the replacement cost, we replace the component rather than repair it. Speed must also be considered. The customer in many cases is more concerned about the length of time required for repair. Availability of replacement parts must then be considered.

Some preventive maintenance checks should be run each time a washer is serviced. Always check the filter screens in the water inlet hoses. Remove any debris that might restrict the water flow. Replace these screens if necessary. Also look for cracked hoses and rusty hose clamps that are a sign of future trouble. Recommend replacement to your customer to avoid callbacks.

Notice the amount of mineral deposits found in the water system when you remove hoses, the agitator, or other parts. This crusty material can cause stuck valves that result in machine overflow. In extreme cases a water softener may be added to the system.

Washer Mechanical Systems

The mechanical components of an automatic washer are associated with the drive mechanism. These pulleys, shafts, gears, and clutches move the basket, tub, and agitator. This achieves both the necessary flexing of the fabrics to remove soil and the high-speed spinning to expel water. The automatic washer mechanical system may at first appear to be very complex. If examined closely, however, it is not difficult to understand.

There are two major functions of the mechanical system: to impart agitating motion to the wash and to give spinning motion to the load. To accomplish this a series of gears, pulleys, shafts, solenoids, and belts are used. Together they convert the mechanical rotation of the electric drive motor to whatever motion is needed to accomplish the wash cycle. Two popular systems are now in use to do this job. One is the *tumbler* system, sometimes referred to as a front-loading machine, that uses a horizontally mounted tub. The second is a *top loader* and uses a vertically mounted tub.

TUMBLERS

Tumbler machines are not as popular in home laundries as they once were. There are many of these machines in

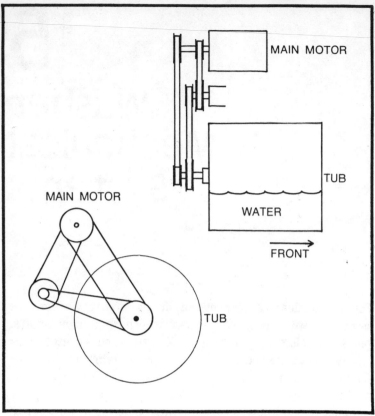

Fig. 5-1. A tumble-washer drive system.

commercial use, however, and some manufacturers still use them for consumer-purchased models. Compared to other methods this system is relatively easy to understand.

Washing action is accomplished by tumbling the clothes in the wash water. This is done by rotating the horizontally mounted tub by means of an electric motor and a pulley and belt arrangement. Water is removed from the clothes by rotating the drum in the same direction but at a much higher speed. The simplified diagram in Fig. 5-1 shows how the tub can be connected to the motor to obtain this operation.

A shaft fixed to the rear of the tub is bearing mounted to the back cabinet of the machine. This shaft supports the rear weight of the tub and its contents. The front of the tub is usually supported by rollers. The motor runs in one direction but has two speeds—one for wash and rinse, another for spin and pump. Mounted on the motor shaft are two pulleys, the main drive pulley and the spin pulley. The connection of these pulleys to the motor shaft is made through a clutch arrangement that mechanically connects and disconnects the pulley and shaft.

The main drive pulley drives a belt that, in turn, drives an idler pulley. A second belt runs from the idler to the main drive pulley on the tub shaft. During wash and rinse operations the clutch on the motor pulley is engaged and this pulley drives the tub through the idler system.

A second pulley is used for the spin function. Another clutch-controlled pulley on the motor shaft pulls the belt that drives a second pulley on the tub shaft. The differences in the gear ratio between the two pulley drives and the motor speed account for the difference in the rotation speed of the tub. The clutches are operated by solenoids and determine which drive chain is used to turn the tub.

How these clutches function can be explained with the aid of Fig. 5-2. With the solenoid deenergized, the spring tension

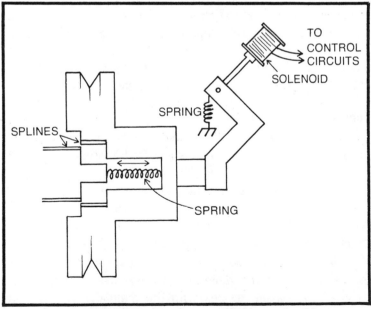

Fig. 5-2. Pulley clutches are controlled by a solenoid.

exerted on the bellcrank lever keeps the pulley disengaged by sliding it outward on the shaft. When energized by the timer circuit the solenoid pulls upward on the bellcrank. This slides the pulley in on the motor shaft and engages the splines on the pulley with the splines on the shaft. The pulley will then rotate with the shaft.

Although there are differences in the pulley and belt arrangements of tumbler-type automatic washers, most operate in a manner similar to that just explained. Should you encounter differences, they should not give you much trouble if you closely examine the drive system and tub-mounting features.

TOP LOADERS

There are several common features among top-loading machines. All must have some means of converting the rotary motion of the drive motor to the back-and-forth motion needed for agitation. They must also have some method of spinning the basket and its load at a high speed to remove the water.

It is important to understand the two most prevalent mechanical-control systems for top-loading machines. Once you understand these two systems, others should give you little trouble. Exact operation, repair, and service information is given in the manufacturer's literature. If you encounter a machine for which you cannot obtain this literature, study the actual machine and note the operation of each component.

Converting Rotary Motion

Most machines use a drive- and sector-gear arrangement to convert motor rotation to back-and-forth motion. Figure 5-3 illustrates the principles employed in most machines. The mechanism uses four gears and one connecting rod to change the motion at the input of the transmission to the motion necessary at the output. In the illustration these are labeled as the pinion, main drive gear, sector gear, and agitator gear.

The drive motor rotates at a constant speed. Assume that the motor drives the pinion in a clockwise direction (as shown). The pinion meshes with the main drive gear and drives it in a counterclockwise direction at the same constant speed. A connecting rod is bearing mounted at both ends, one end to the sector gear and one end to the drive gear. As the drive gear turns about its shaft, the end of the connecting rod attached to the drive gear is rotated about the drive gear shaft.

The sector gear is also bearing mounted to its shaft. This gear could be compared to a lever with gear teeth at one end and a connecting rod at the other. Its name comes from the fact that it is not a complete gear (360°) but covers only a sector (or portion) of a complete circle. The sector gear meshes with the agitator gear on the agitator shaft. The oscillations of the sector gear will cause the agitator gear to turn first in one direction, then reverse itself and turn in the opposite direction.

In Fig. 5-3A the connecting-rod mounting-bearing pin is at the high point of its travel and positions the sector gear at its mid position. As the pinion gear turns the main drive gear counterclockwise, the lower arm of the sector gear is pushed to the left. This moves the sector gear to the right. Since the agitator gear meshes with the sector gear the agitator gear and shaft turns counterclockwise.

In Fig. 5-3B the drive gear has made one-fourth of a complete revolution and has moved the sector gear to its extreme right position. As the drive gear continues to turn counterclockwise, the sector gear movement is reversed. When the pin has completed one-half of its revolution (Fig. 5-3C) the agitator shaft is at its original position but moving in the opposite direction. As the connecting rod moves right, the sector gear moves left and the rotation of the agitator shaft is reversed and turns clockwise. This continues until the gear reaches the point where it has completed three-fourths of its revolution as shown in Fig. 5-3D. At this point the agitator gear would again be reversed and turn clockwise.

Separate Clutch and Transmission

Some automatic washers employ a drive system that includes an enclosed transmission and an external clutch. Most machines of this type house inside the gear case the mechanical components that convert rotary motion to oscillating motion. The transmission and its associated

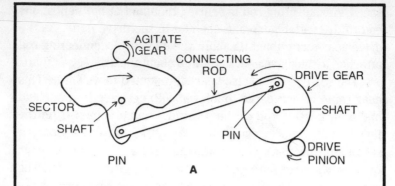

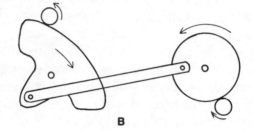

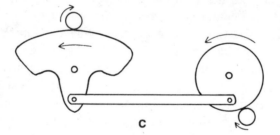

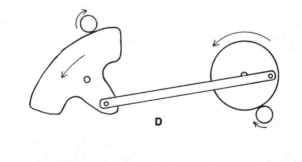

Fig. 5-3. The sector drive gear provides the necessary back-and-forth motion.

components perform the agitating function. Spin operation is controlled by an external clutch.

The operation of this system can be understood with the aid of Fig. 5-4.

The main drive motor is connected to the agitate and spin mechanisms through a series of pulleys, three in this case. Power from the motor is transmitted by a belt to the spin drive pulley and agitate drive pulley. Both pulleys turn whenever power is applied to the motor. The timer will determine which function is to be performed at any given instant by energizing either a spin solenoid, which engages the spin clutch, or an agitate solenoid, which engages the gears within the transmission. Both solenoids are never energized at the same time.

A drive shaft from the transmission enters the tub through a seal. Some tubs have a built-in drive tube through which the drive shaft runs. On others, only a seal at the bottom of the tub prevents water from escaping. When the agitate solenoid is energized by the timer, a gear mesh is made inside the transmission. The drive motor turns, causing the agitate drive pulley to rotate. This drive pulley drives a pinion gear as shown in Fig. 5-3. Oscillating motion is applied to the agitator

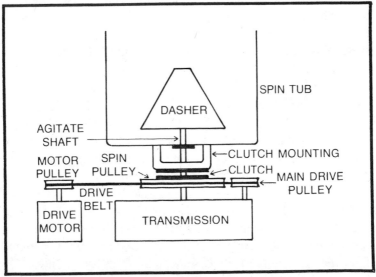

Fig. 5-4. An example of an external clutch.

shaft by a sector gear. We now have the familiar back-and-forth motion of the dasher.

During spin periods the agitate solenoid is deenergized and power is applied to the spin solenoid. Spin tension inside the transmission causes the gear mesh at the pinion to be broken. This allows the agitate drive pulley to "freewheel." At the same time, the spin solenoid is energized bringing the upper and lower clutch surfaces together. Now the power from the drive pulley spins the entire basket. During agitation the spin drive-pulley freewheels about the agitator drive shaft.

This system uses a single-direction motor. The drive pulleys are turned whenever the motor supplies power and always turn in the same direction.

Internal Clutch

Some mechanical systems use a transmission designed around an internal spring clutch to transmit mechanical energy to the dasher and spin basket. A sketch of such a system is shown in Fig. 5-5. The motor transmits mechanical energy to the transmission by means of a belt and pulley drive chain. In this system a bidirectional main drive motor is employed. During the wash/agitate and rinse/agitate cycles the motor turns in one direction (counterclockwise). During spin and pumpout functions the motor direction is reversed. A spring-friction clutch is used to automatically shift the transmission into the mode corresponding to the direction of motor rotation.

Later in this chapter we will examine the workings of the clutch in more detail, but for now assume the clutch operates as stated. As the motor turns in the counterclockwise direction the clutch binds the main drive shaft to the drive pinion as shown in Fig. 5-5. The pinion then drives the main drive gear. A sector gear arrangement, similar to that of the previous system, converts the constant rotation of the main drive gear to the back-and-forth motion necessary for agitation. Seals and bearings permit the agitator shaft to enter into the bottom of the tub and prevent any water leaks around the agitator shaft. Likewise, the agitator shaft and main drive shaft enter the transmission through oil seals.

During spin and pumpout cycles the motor direction is reversed and the motor rotates clockwise. Motion is again imparted to the main drive shaft, but in the opposite direction. The spring-friction clutch now applies enough torque to the pinion to cause the ratchet pawl to engage. This locks the transmission to the case and the entire case is turned by the main drive pulley.

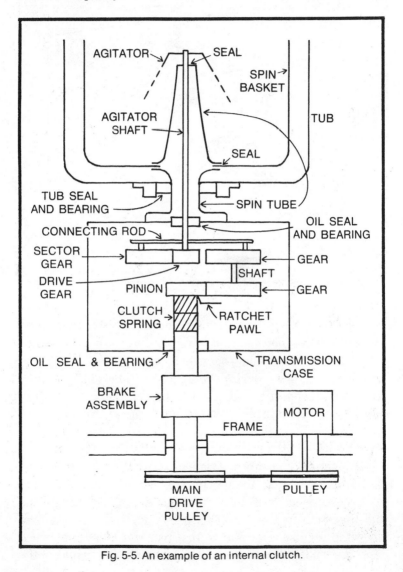

Fig. 5-5. An example of an internal clutch.

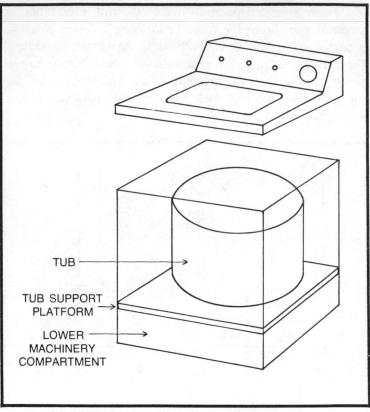

Fig. 5-6. Tub mounting is similar for many machines.

The weight of the transmission and its associated components is supported by a lower bearing in the framework of the cabinet. A solenoid-operated brake is used to hold the spin basket in place during the agitate period. This device is electrically connected to a lid switch that applies the brake to the drive system if the lid is raised during any spin period.

TUB CONSTRUCTION AND MOUNTING

The tub components of a typical automatic washer consist of the main water tub, the spin basket, and seals and gaskets at the necessary entrance holes in the tub. Actual mounting of the tub assembly to the cabinet varies. However, the method shown in Fig. 5-6 is similar to that found in many machines. Here the tub is supported on a platform in the lower part of the machine cabinet. The tub is bolted to the platform to rigidly attach it to the framework. Variations of this design omit the platform and use several pieces of channel iron to form a suitable mounting framework.

The components of a typical tub assembly are shown in Fig. 5-7. A main water tub encloses the entire assembly. Most

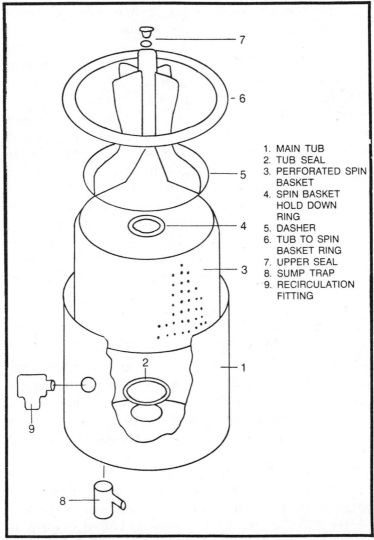

1. MAIN TUB
2. TUB SEAL
3. PERFORATED SPIN BASKET
4. SPIN BASKET HOLD DOWN RING
5. DASHER
6. TUB TO SPIN BASKET RING
7. UPPER SEAL
8. SUMP TRAP
9. RECIRCULATION FITTING

Fig. 5-7. The components of a typical tub assembly includes several parts contained within the main tub.

manufacturers use a tub designed with as few penetrations as possible to reduce the chances of leaks. In this example only three penetrations are used. One at the bottom of the tub provides entry for the dasher and spin drive components. A second opening is the trap where water is drawn off. The third fitting, higher on the side of the tub, is a connection for the recirculation hose.

A perforated basket fits inside the main tub. This basket supports the clothes in the water during agitation and also revolves at a high speed to remove excess water from the clothes during the spin cycle. Seals are used at the bottom of the tub and at the bottom of the spin basket. Tub seals prevent water leakage from the main tub while the gasket and ring attached to the spin basket connect it to the centerpost assembly. Although no water seal is provided for the spin basket, the gasket allows the basket to be tightly connected to the centerpost without damaging the finish.

A bolt at the top of the dasher secures it to the agitator post. The agitator post is driven by the agitator drive shaft inside the spin tub. Separate mounting of the spin basket to the spin tube allows the basket and agitator to be driven by different drive shafts. The large plastic molding at the top of the tub forms a cover between the spin basket and the water tub. This prevents small articles of clothing from falling between the water tub and spin basket.

TRANSMISSION OPERATION

There are many physical similarities among the components and operations of different automatic washer transmissions. This is not to say that all are identical or that parts are interchangeable. It does mean, however, that the principles discussed here are applicable to most washers.

Solenoid-Controlled

The operation principles of a solenoid-controlled transmission can be illustrated with the exploded view in Fig. 5-8. Back-and-forth agitation is achieved with a sector gear arrangement as previously explained. The main drive gear is turned by the movement of the main drive pulley. A connecting rod moves the sector gear back and forth. No power is applied to the solenoid during the agitation of wash and rinse cycles. Spring pressure keeps the agitator gear meshed with the sector gear. Power applied to the solenoid during spin cycles causes the solenoid to exert an upward force on the agitator fork. This lifts the agitator gear above the sector gear and stops the back-and-forth motion of the agitator shaft.

Spin operation in this washer also uses a solenoid control and is illustrated in Fig. 5-9. This unit mounts on top of the transmission assembly shown in Fig. 5-8 in a manner that extends the agitator shaft upward through the spin mechanism. A spin drive pulley is turned by the drive belt when the main drive motor is running. The spin disc engages the drive pulley only during the spin operation, however. During agitation the solenoid bar holds the yoke up and prevents the spin disc from touching the spin drive pulley. A brake disc on top of the yoke pushes against the spin brake

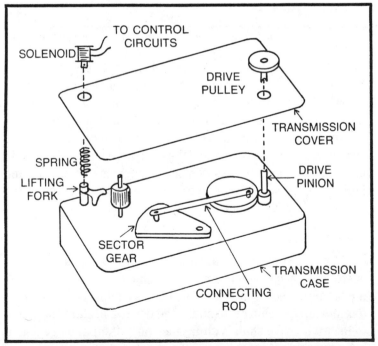

Fig. 5-8. Exploded view of a solenoid-controlled transmission.

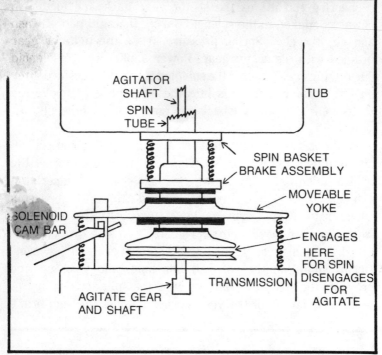

Fig. 5-9. The spinning action is also solenoid controlled.

assembly during agitation. This holds the spin tub stationary. During spin cycles the agitate gear is disengaged, as explained in conjunction with Fig. 5-8, and the solenoid in Fig. 5-9 moves the solenoid bar outward. A detent in the solenoid bar allows the yoke spring to pull the yoke down. This releases the brake and, at the same time, engages the spin drive disc with the spin pulley. The spin tube extends up through the main tub and attaches to the spin basket. Spin disc movement is transmitted to the spin tub by the tube.

Self-Contained

The self-contained transmission is designed around the operation of a spring-tension friction clutch. Two views of such a clutch are shown in Fig. 5-10. The assembly is made up of four parts: the main drive shaft, the drive shaft hub (attached to the main drive shaft with a taper pin), the pinion gear and hub assembly (one piece), and the clutch spring.

140

Only the spring makes mechanical connection between the gear and shaft hubs. The gear is free to rotate about the shaft even though the shaft extends up through the gear. At the point where the shaft extends through the gear, a ring key holds the gear on the shaft. The ring key also holds a pawl mechanism to the shaft. The spring fits tightly over both hubs.

If you view the assembly from the drive shaft end, and the spring is wound in a clockwise direction as shown, turning the shaft counterclockwise would make the spring coils bind together. This would increase the spring tension on the hubs and cause them to act as a solid shaft. Counterclockwise motion of the shaft would then bind the hubs together and the pinion gear would rotate counterclockwise.

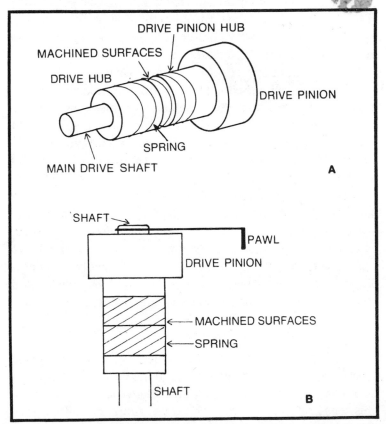

Fig. 5-10. Two views of a spring-tension friction clutch used in a self-contained transmission.

When the shaft is turned clockwise, in this example, the force applied to the spring causes the spring to unwind. Since the spring cannot actually unwind, the reverse tension on the coils allows the hubs to slip inside the springs. Torque cannot be exerted on the pinion gear since the hubs tend to slip inside the springs, but the pawl is driven by the extension of the drive shaft through the pinion gear. The pawl is driven only part of a complete revolution until it contacts the main drive gear. The transmission locks when the pawl engages the main drive gear. The clockwise torque on the drive shaft then causes the entire transmission, case and all, to turn clockwise.

This type of transmission is suspended by two bearings. The lower end of the assembly is mounted on a heavy thrust bearing while a similar bearing holds the upper end in place. The drive shaft extends through the lower bearing and attaches to the main drive pulley. The agitator shaft and spin tube extend upward through the upper spin bearing and attach to the agitator and spin basket. This is shown in Fig. 5-11.

The brake hub is bolted to the bottom of the transmission case. The lower part of this assembly is a machined surface with a brakeband fitted around it. The brakeband fits tight enough to apply friction to the hub but not so tight as to be bound to the hub. In other words, it can slip—slightly—around the hub. As shown in Fig. 5-11 a latch lever is held onto the brakeband by spring tension. This holds the transmission case and spin basket rigid during agitation cycles. During spin cycles the timer energizes the brake solenoid. This lifts the latch from the brakeband and allows the transmission case to spin. The electrical circuit for the solenoid usually includes a lid switch in series with the solenoid. Raising the lid will interrupt the circuit and stop the spinning action.

SERVICING

Most manufacturers go to great lengths in designing and constructing transmission systems to insure a relatively long and trouble-free life. Nearly all automatic washer transmissions are backed by a five-year warranty. Considering the heavy use this component receives, this shows great confidence on the part of the manufacturers. These units are

142

assembled, adjusted, and lubricated at the factory. In many cases they are never opened again for the life of the machine. Although these are very reliable mechanisms they will occasionally need service, however.

Lubrication

Automatic washer transmissions are lubricated by a grease bath. The transmission case is sealed and filled with gear grease. Most manufacturers specify SAE 90 weight gear grease. If it becomes necessary to add grease to the transmission, consult the manufacturer's service literature. On many models it will be necessary to completely remove the transmission from the machine.

Adjustments

In early models, transmission adjustments to the gear mesh had to be made internally. Few modern transmissions

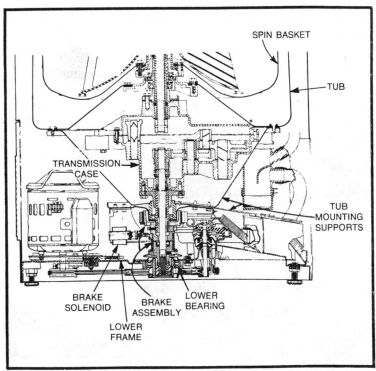

Fig. 5-11. Transmission mounting using bearings.

have internal adjustments, however. Most machines do have mechanical adjustments that must be made occasionally, but they are mostly outside the transmission case. Always refer to the manufacturer's service literature for instructions when making these adjustments.

Brake and clutch mechanisms outside the transmission case will sometimes need adjusting. These adjustments should also be made according to manufacturer's instructions. To insure proper operation and customer satisfaction, make these adjustments as close to specifications as possible. Take precautions to prevent dirt or oil from contacting clutch or brake linings, since this will interfere with friction coupling existing at such surfaces.

Disassembly and Assembly

These procedures will vary from make to make and model to model. Literature available from the manufacturer will give step-by-step details. Here we will give you a general idea of what is necessary to remove the transmission from a typical machine.

First remove the top cover and console panel. On many machines they can be removed as a single unit.

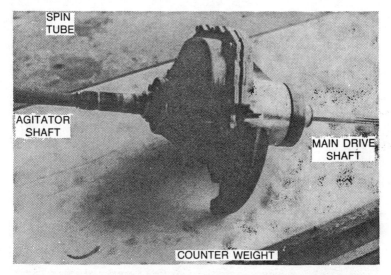

SPIN TUBE

AGITATOR SHAFT

MAIN DRIVE SHAFT

COUNTER WEIGHT

Fig. 5-12. A transmission assembly showing the topside components.

Quick-disconnect plugs are usually used between the controls on the console and the lower unit components. Spring clips hold the main tub cover in place. To remove this cover it is often necessary to remove the water inlet plume and laundry agent dispensers.

The agitator is usually held in place by a capscrew on the agitator post. After the agitator is removed, the agitator drive block that mates with the agitator must be removed. The spin basket is held to its drive component by a holddown ring and several bolts. Once the holddown ring has been removed the spin basket may be lifted out. Next, the main water tub is removed in a similar manner. Now the transmission can be lifted out through the top of the machine.

On other models, the transmission is removed through the bottom of the machine by first disconnecting the tub and removing the bolts that hold the transmission to the frame.

Like transmission removal, transmission-case disassembly will vary from machine to machine. The photographs in Fig. 5-12 through Fig. 5-15 illustrate a typical disassembly operation. Reassembly is the reverse order.

MECHANICAL PROBLEMS

Mechanical problems in automatic washers may come in one of several ways. Worn bearings, bushings, or gears may show up as noisy operation, sluggishness, or complete binding

Fig. 5-13. A transmission assembly showing the underside components.

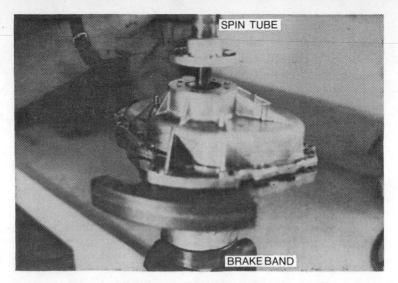

Fig. 5-14. Removing the spin tube.

of the mechanical components affected. Defective seals can be revealed by oil in the wash water or leaks underneath the washer. Troubleshooting the mechanical systems is a simple process in most cases. Usually, these troubles can be located by close observation of the machine's operating cycle.

Some mechanical problems may be corrected in the customer's home. If it is necessary to open the transmission, however, it is recommended that the transmission be taken to the shop. This will not only make it easier to work with the transmission, but there will be less chance of damage to the customer's home. Some larger shops specializing in repair of only a few makes of machines keep a stock of rebuilt transmissions. This lessens the downtime of the machine and increases customer satisfaction. Always check the warranty expiration date when servicing mechanical components. As mentioned previously, mechanical components are usually covered with a long warranty and you may be able to save the customer a considerable amount of money. Most servicemen refer customers with warranty machines to shops authorized to do warranty work on those machines. This may lose you one call, but if you can save the customer some money you'll earn a friend and accomplish a bit of PR for your shop.

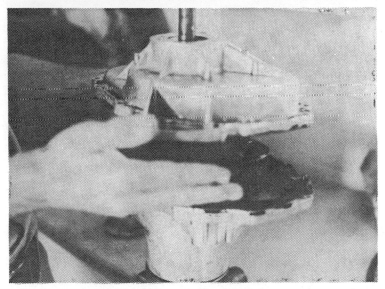
Fig. 5-15. Removing the transmission cover.

During reassembly be sure that the mechanical components are clean and free of dirt and grit. Clean all gears with solvent and dry them with compressed air. When replacing gaskets, use a good-grade gasket sealer and make sure the gaskets are flat and smooth. Seals should be replaced with care. Seal installation tools that reduce the difficulty of seal removal and installation are available from your parts jobber or from the manufacturer.

6

Electrical Systems

Of the three major systems comprising an automatic washer, the electrical system varies less from machine to machine than any other. Solenoids convert electrical energy to the mechanical motion needed to open and close valves, shift gears, and engage clutches. Timers set the pace for the entire machine operation. These timers are multicontact, multipole switches driven by a small electric motor. Depending on the position of the switch, various solenoids, switches, valves, and even the drive motor is set to the operating condition corresponding to the wash cycle. Solenoids and their operation were explained in an earlier chapter of this book. Here the timer and the overall electrical operation of an automatic washer will be discussed.

MACHINE TIMING

The timer is the brain of the automatic washer. Its ability to keep track of time and sequence makes it seem like a very complex component. You will find it can be divided into two separate devices: the timer switch and the timer motor. The switch, as mentioned previously, compares to a multicontact rotary switch. The major difference is the manner in which the contacts are opened and closed. A small electric motor drives

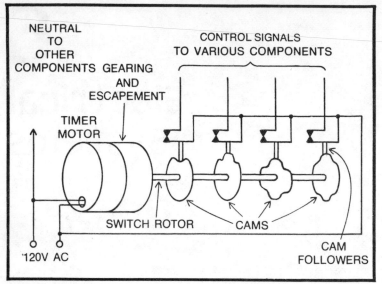

Fig. 6-1. A simplified diagram of a timer. Notice the set of cams driven by the timer motor.

an escapement similar to that in a clock. The escapement allows the motor to drive the timer switch in discrete steps, instead of continuously rotating the switch. In other words, the switch moves around in very small increments instead of turning smoothly.

Timers

A simplified diagram of a timer is shown in Fig. 6-1. A series of cams attached to the switch rotor are driven by the timer motor through the escapement. Cam followers are held against the cams by spring tension. As the cam rotates to where the cam follower rides a high spot in the cam, those contacts are closed. When the cam moves to where the cam follower is at a low point, the switch contacts open (see Fig. 6-2). The cams are cut so that contacts and solenoids are energized at the correct time to insure proper operation of the wash cycle.

Cycles

Timers control machine operation through several cycles. Most machines sold today have at least two different wash cycles. Some have as many as six or eight. No matter how many cycles are featured on the machine, or how many automated operations are accomplished during each cycle, the time that these operations occur is controlled by a timer similar to the ones discussed in this text.

A close study of the timer dial on any machine will reveal many of its operation features. If you are called to service a machine with which you are not familiar, your first step in servicing should be to examine the timer dial. Note the number of cycles available and the different operations that make up each cycle. Figure 6-3 illustrates some of the features you will see in a typical two-cycle machine.

First, you should notice that there are two separate cycles of operation—a *regular* wash cycle and a *delicate* cycle. The regular wash cycle is for washable fabrics such as colorfast cottons. The delicate cycle is primarily for more delicate materials, such as knits and noncolorfast fabrics. Each of these cycles is divided into smaller increments in which specific operations are carried out. Typically, the regular cycle will encompass 30—40 minutes, while the delicate cycle will be shorter. Similarly, each of the incremental operations is usually shorter in the delicate cycle. This results in a cycle that exposes delicate fabrics to a gentler wash than they would receive in the regular cycle.

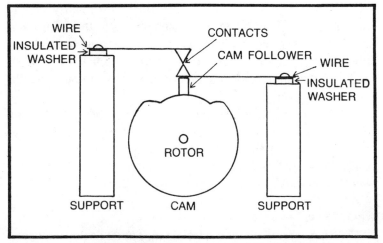

Fig. 6-2. Cam-switch arrangement of a timer.

Fig. 6-3. A typical timer dial for a two-cycle washer.

Each timer has a definite number of cycles through which it controls machine operations. And each cycle has definite steps. Learning to recognize these steps, and what operations should be completed during each step, is a major part of learning to troubleshoot automatic washers.

Most manufacturers attach a wiring diagram and timing chart to each machine. These diagrams are also included in the manufacturer's service literature. To efficiently repair automatic washers, you not only need copies of these for each machine you service, but you also need to be able to read and understand what the diagrams tell you. If you know how to read the diagrams, you know exactly what the machine should be doing and exactly when it is to do it.

The simplified timing chart in Fig. 6-4 shows the timer sequence employed with the dial shown in Fig. 6-3. Only one cycle is shown in this particular example. The same cycle would be used for delicate fabrics except that each period would be shorter.

Water-Level Controls

Timing operations on most machines start at the end of the fill period as shown in the diagram. Some earlier models have timed periods during which the machine is filled. Most modern machines have done away with this and placed the timer motor in the circuit so that it doesn't start until the machine is full.

Once the water-level control senses that the tub has filled with water, the water inlet valves are closed and the main drive motor is started. The machine continues to agitate the clothes for about 14 minutes during a regular cycle and about 8 minutes in the gentle cycle. Machines with automatic bleach and detergent dispensers release their products into the wash water during the agitation cycle.

At the end of agitation, the main drive motor is stopped and there is a short pause. This gives the drive motor a chance to coast to a stop. Usually, about one minute is allowed for this. The timer now engages the proper contacts to start the drive motor for the pumpout operation. Many machines use reversible motors driven in one direction for the pump and spin operations and in the opposite direction for agitation. Other machines use a gear mechanism that changes the gearing arrangements between agitate and spin/pump times.

The timer would now move from the *stop* position to the *pump/spin* position. Mechanical devices shift into the pump-

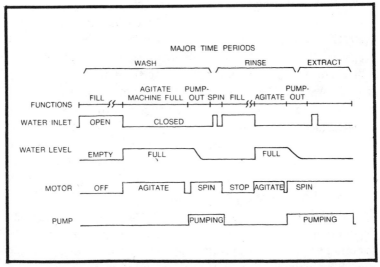

Fig. 6-4. A simplified washer timing diagram whose cycles correspond with the dial in Fig. 6-3.

out phase and water is pumped from the machine. When the water level in the tub falls to a specified level, the basket starts spinning to remove water from the clothes. Most machines use a spray/spin feature where the water inlet valve is intermittently opened to allow water into the tub during the spin cycle. The spraying action on spinning clothes removes soap scum left behind by receding water. Typically, this spin cycle runs about 8 minutes.

A second short pause is encountered at the end of the spin period. This allows the load to cool for a short time as well as providing a short delay for shifting modes of operation. When energized by the timer, the machine will begin to fill with rinse water. The main drive motor is not started until the water-level switch indicates a full tub.

After filling with rinse water, the agitate mechanisms are again engaged. This works the clothes back and forth through the rinse water. If the machine has automatic dispensers for rinsing agents, these devices would be energized during this cycle. This agitate operation lasts for about 6 or 8 minutes during most regular cycles, and 5 minutes or so during the delicate cycles.

Electrical Operation

Two illustrations will be used to explain the electrical operation of an automatic washer and the timing of various steps. Figure 6-5 is the wiring diagram of a two-cycle washer, while Fig. 6-6 is the timing chart for the same machine. In actual practice you can usually find both these diagrams attached to the machine. Understanding how to use them is the key to understanding the machine's operation.

Let's start with Fig. 6-5. The common side of the 120V supply is brought into the electrical system through the push-pull on/off switch. This switch is attached to the shaft of the timer knob. To start the operating cycle the switch must be pulled out. The common side of the line is then brought to terminal 1 of the main drive motor, terminal 1 of the timer motor, terminal 1 of the brake solenoid, and terminal 1 of the water inlet valve. Notice that terminal 1 of the water inlet valve is common to both the hot and cold solenoids.

The hot side of the line is routed to pin 21 of the timer through the overload switch. This is an automatic thermal-overload protector. Whenever current or motor temperature is too high, the overload opens the circuit and removes all power. When the motor current or temperature drops to a safe level, the overload switch will reset and restore power to the circuit.

Now refer to the timing chart in Fig. 6-6. Note how the chart is laid out. This is typical of many machines. Across the top, the cams that control the various functions are listed. Ten cams, A through J, are used. Cam A is used to control water temperature. This switch signals the water-temperature

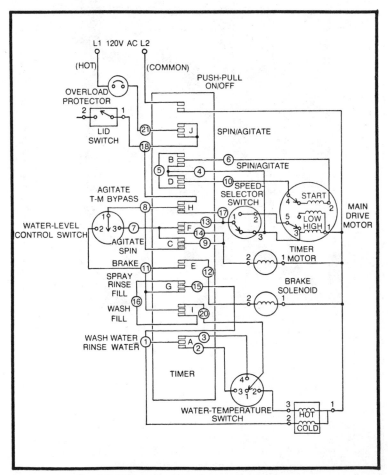

Fig. 6-5. Wiring diagram of a two-cycle washer.

155

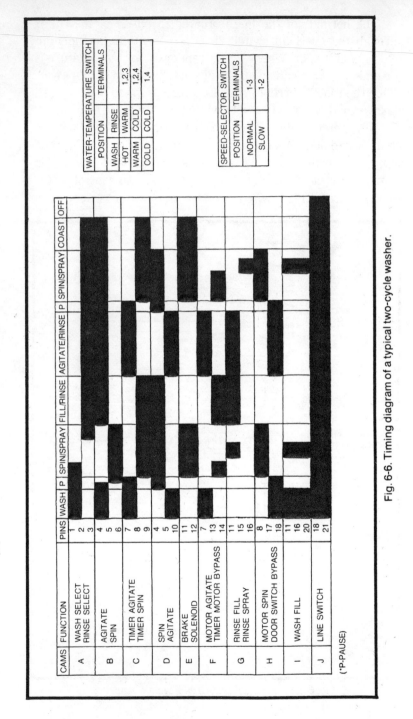

Fig. 6-6. Timing diagram of a typical two-cycle washer.

(*P-PAUSE)

156

switch as to whether the machine is in a wash or rinse period. Cam B designates agitate or spin functions in a similar manner. As you continue to read across the chart, the function controlled by each cam is listed. The last cam, J, is used by the timer to turn off the machine at the end of the cycle.

Both cycles are shown in the diagram. The top of the chart lists the operations for the regular cycle, while the lower portion shows gentle cycle functions. The specific periods of operation are listed to the right of the cycles. Water-temperature and speed-selector switches are shown below the general timing chart. The speed selector in this case is a separate cam positioned to close terminals 1 and 3 during the normal wash cycle and terminals 1 and 2 during the delicate cycle.

The water-temperature switch is a separate control positioned by the machine operator. Three temperature selections are provided: hot water wash/warm water rinse; warm water wash/cold water rinse; cold water wash/cold water rinse. On some machines this feature is built into the timer, and water temperature is selected on the basis of the cycle. In this example we use a separate switch.

To operate this machine in the normal cycle, the operator selects the desired temperature, loads the machine with clothes and detergent, puts the timer to the start of the regular cycle, and pulls out the knob. The timing diagram shows cam A is now in position to select wash water. Each of the other cam positions are shown to the right.

For the remainder of this discussion it will be necessary to refer to both Fig. 6-5 and Fig. 6-6.

As soon as the operator starts the machine, the common side of the 120V supply is connected as previously explained. At the same time, the hot side of the line is connected through the overload switch to pin 21 of cam J. This is sometimes called the on/off cam since it provides a means for the timer to turn off the machine. In this *off* position, the switch is in the position shown. In the normal cycle pin 21 is made with the upper contact, while the delicate cycle closes pin 21 with the lower contact. Regardless of which cycle is selected, cam J applies power to pin 18 of the timer. At the end of the cycle, cam J removes that power.

Power is routed from pin 18 through parallel circuits to the water-level switch. One of these paths is through the lid switch, the other through cam H. Pin 18 of cam J is common with pin 8 of cam H. At this time the machine is in the wash/fill cycle. Therefore, pins 18 and 8 of cam H are closed. Power is applied to terminal 1 of the water-level switch, even with the lid open. Since the tub is empty, terminals 1 and 2 of the water-level control are closed, applying power to pin 11 common to cams E, G, and I.

According to the timer chart (Fig. 6-6), cam E holds its associated contacts open during the wash/fill function. Therefore, these contacts are not significant at this time. Cam G, as shown by the timing chart, is also inactive at this time. Cam I is active, however, and closes pins 11 and 20 applying power to terminal 1 of the water-temperature switch.

Assuming that cold water has been selected for the wash operations, terminals 1 and 4 of the temperature switch are closed. This routes power back to the timer at pin 3 of cam A. During the wash/fill function, cam A closes pins 1 and 3 supplying power to the cold water solenoid. Should hot water be selected, terminals 1, 2, and 3 of the water-temperature switch are closed. Power is applied to the hot water solenoid through terminals 1 and 2 of this switch. Terminal 3 of the water-temperature switch would supply power to pin 2 of cam A. This cam holds pins 1 and 3 closed and pin 2 open at this time, however.

When warm wash water is selected, terminals 1, 2, and 4 of the water-temperature switch are made. Terminals 1 and 2 energize the hot water solenoid. Power from terminal 4 routes back through pins 3 to 1 of cam A to turn on the cold water.

Although the common side of the power source is connected to the timer motor at this time, the hot side is held open. According to the timer chart, cam C is closed from pins 9 to 7, and cam F is closed from pins 13 to 7. This means that before the timer motor can start, or power can be applied to the main drive motor, terminal 3 of the water-level control must be hot.

The brake solenoid controls a brake mechanism used to prevent the spin basket from turning during agitation. During spin cycles the solenoid is energized. This lifts the brake and allows the spin basket to turn. As we have already pointed out, pins 11 and 12 of cam E are now open and the brake is applied.

As you can see, the machine will remain in this condition until the water-level switch indicates that the water has reached the proper level. Many machines have a variable control to select the water level. This control functions in the same manner as the one shown here. Mechanical features of this switch vary the spring tension, allowing for water-level selection. This added feature does not affect the electrical circuits and, from a wiring diagram, you cannot distinguish between fixed and adjustable water-level controls.

The water-level switch closes terminals 1 and 3 and opens terminals 1 and 2 when the machine fills. This applies power to pin 7 of cams F and C. Cam C has pins 7 to 9 closed, starting the timer motor. At cam F, pins 13 and 7 are closed and power is applied to terminal 1 or the speed-selector switch. Speed is controlled by the cycle selected. During the normal wash, terminals 1 and 3 of the speed-selector switch are closed, placing power on the high-speed winding of the motor. For delicate cycles terminals 1 and 2 of the speed-selector switch are closed and power is applied to terminal 5 of the motor. Notice that terminal 3 of the speed-selector switch is connected to pin 4 of cams B and D. These are the spin and agitate cams. During agitation cycles, cam B closes pins 4 and 6 while cam D closes pins 5 and 10. To control spin and agitation, these two cams are used to determine the direction of rotation for the main drive motor. From an earlier chapter you learned that to reverse the direction of rotation of an induction motor it is necessary to reverse the polarity of the motor-start winding. That is accomplished with cams B and D. Terminal 4 of the motor-start winding is connected to pin 10 of cam D, while terminal 2 of the motor-start winding is connected to pin 6 of cam B. During agitation cycles, the start winding is connected from terminal 2 of the main drive motor to pin 6 of cam B, to pin 5, and to the common side of the line. Terminal 4 is connected to pin 10 of cam D, to pin 4 of the cam, and to the hot side at terminal 3 of the speed-selector cam.

Notice that the connections to terminals 2 and 4 of the start winding will be reversed when the cams shift to a spin function. Also note the internal centrifugal switch of the motor. When the motor is at *rest* the switch contacts are made as shown. When the motor reaches operating speed, the switch opens terminal 4 of the start winding and terminal 3 and 5 of the run circuit. At the same time, pins 5 and 6 at cam B are closed. This means the motor is always started in the high-speed (regular cycle) mode. If low speed (delicate cycle) is selected, terminals 1 and 2 of the speed-selector switch are closed, applying power from pin 13 of cam F to terminal 5 of the motor. When the motor reaches operating speed, the centrifugal switch will remove power from the high-speed winding and apply power to the low-speed circuit.

This particular machine has a 10-minute wash time during the regular cycle. Notice on the timing chart that all mechanisms remain in their established conditions until the wash period terminates at the 10-minute mark. At that time, cams D, F, and I change their contact positions. Cam D opens pins 5 to 10 and closes pins 10 to 4. This shifts terminal 4 of the motor-start winding from the common side of the supply to the hot side. Cam F opens pins 13 to 7, removing power from the speed-selector switch and the main drive motor. Cam I opens pins 11 to 20, removing power from terminal 1 of the water-temperature switch. A short pause is allowed for these actions to take place. After about one minute, cam B opens pins 4 to 6 and, at the same time, closes pins 6 to 5. This action connects terminal 2 of the motor-start winding to the hot side of the line at terminal 3 of the speed-selector switch. Reversing the polarity of the start winding is now complete.

At the end of the pause, cam E closes the pins to the brake solenoid, unlocking the brake from the spin basket. This starts the spin/spray operation, which lasts for approximately six minutes. Reading across the timing chart, you can see that cam A has both sets of pins open. Cams B, C, and D are in the spin position and cam E has removed the brake. Cam F, during the first few moments of the period, closes pins 7 and 14. This insures that power continues to the timer motor when cam C is changing the position of its pins. Notice that this happens each time cam C changes position. Cam G closes pins 15 and 16 for a short period. This ties terminal 2 of the cold water solenoid to the hot side of the line at terminal 2 of the water-level switch. When the water is down to a level that causes the water-level control to switch from the *full* position to the *empty* position, the cold water solenoid is activated and a spray of cold water enters the tub.

Upon completion of the spin/spray cycle, cams B, C, and D shift back to the agitate function and the fill circuit is energized to fill the machine with rinse water. Cam A now connects pins 1 and 2, and the water-temperature switch will control the water inlet valve by energizing either the cold water solenoid or both hot and cold water solenoids, depending on whether warm or cold water has been selected for rinsing.

Most machines have a built-in safety feature that prevents the spin operation from taking place when the lid is open. Looking at Fig. 6-5, you will notice the lid switch in the circuit between pin 18 of cam J and the water-level switch. Should the lid be opened during the spin function, the lid switch will remove power from terminal 1 of the water-level control. This removes power from the main drive motor and the brake solenoid. The main drive motor stops and the brake is released, bringing the machine to a sudden stop.

The remainder of this cycle is similar to the operations already described. The cams continue to shift the switches back and forth to control the various solenoids. Delicate cycles are completed almost identically to regular cycles except for the time allowed for each function and the speed of the main drive motor. Always study the timing charts and wiring diagrams of new machines you service. Notice changes or variations with which you are not familiar. Trace out the circuit and refer to the timing chart to determine how these operations are accomplished.

TESTING ELECTRICAL COMPONENTS

Voltage and resistance measurements can be used to check any electrical component of an automatic washer. Test lamps can be made or purchased to test these devices. We recommend the use of a VOM, however. Test lamps require "hot" leads that can damage the machine and expose the technician to electrical shock. In all instances in this book we will describe tests that take voltage and resistance measurements using a VOM. When possible, resistance measurements are preferred over voltage checks. Resistance checks require no outside power source, thus removing the danger of shock.

Wiring Checks

The wiring hookups inside the washer will give the technician very few problems, mainly because failures seldom occur in this area. Usually, failures here are due to loose connections, chaffed or broken wiring, or improper wiring caused by inexperienced personnel attempting to make repairs. Symptoms of wiring failures are too varied to list, and range anywhere from opens and shorts to shocks.

Many failures encountered in the wiring can be located by simply going through the wiring harness and looking for chaffed spots, and loose or broken leads. If you suspect the wiring, use a diagram and check the wiring from point to point. Check continuity from point to point. These tests will usually reveal any fault in the wiring. Voltage tests can be made; however, power must be applied to the machine and this creates a shock hazard.

Solenoids

Solenoids can be tested using either voltage or resistance measurements. As an example, assume that the machine shown in Fig. 6-5 fails to fill with cold water. A voltmeter can be connected across the cold water solenoid and the timer (turned to one of the fill positions). Select cold water wash and rinse with the water-temperature switch and observe the meter for voltage at the solenoid. Repeat the test for each fill position of the timer and each position of the water-temperature switch that activates the cold water solenoid (warm and cold in both wash and rinse cycles). If voltage is present during any of these cycles, and cold water still does not enter the tub, then the solenoid is definitely at fault. Should you not be able to find any voltage at the solenoid, look for trouble in the timer or water-temperature control.

Resistance measurements are also effective for checking the electrical properties of the solenoid. These checks will not reveal difficulties such as sticky mechanical parts, however.

To make resistance checks, remove all power from the machine and set up an ohmmeter for R × 10 operation. Connect the meter across the solenoid. Make sure there are no parallel circuits to interfere with the meter reading. The exact resistance of the solenoid may vary, but in most cases it runs several hundred ohms. An open coil will be indicated by an infinite meter reading, a shorted coil by a very low reading.

Water-Temperature Switches

As with other components, voltage tests of water-temperature switches must be made with the switch installed in the machine and power applied. To test the water-temperature switch in Fig. 6-5, connect one end of the voltmeter to the common terminal of the water inlet valve (terminal 1). Set the timer to the *wash/fill* position, and set the water-temperature control to *cold wash*. Touch the other probe of the voltmeter to terminal 1 of the water-temperature switch. Voltage should be present here, indicating that the timer is applying voltage at the proper time. Move the probe to terminal 4 of the switch. Voltage here indicates the switch is capable of selecting cold water. Change the switch to *warm wash* and check for voltage at terminals 2 and 4 of the switch. If voltage is present here the warm water contacts are working. Hot water selection can be checked by moving the water-temperature switch to *hot* and checking for voltage at terminals 2 and 3.

Ohmmeter tests can be made by removing all power from the machine and connecting one probe to terminal 1 of the water-temperature switch. When set for cold water wash, the resistance between terminals 1 and 4 should be zero and between terminals 1 and 2 and terminals 1 and 3, infinite. Selecting *hot water* should give zero resistance between terminals 1, 2, and 3 and infinite resistance between 4 and any of the others. When *warm wash* is selected, zero resistance should be measured between terminals 1, 2, and 4 and infinite resistance between these three and terminal 3. Should the switch fail any of these tests, replacement is necessary.

Water-Level Control

Voltage measurements to check the water-level control are made by first setting the timer to the *wash/fill* position.

Attach one probe of the voltmeter to the common connection of the water inlet valve (terminal 1). With power applied to the machine, touch the other probe to terminal 1 of the water-level control. Lack of voltage here indicates trouble elsewhere in the electrical system. With voltage present at terminal 1 of the water-level control, and the machine in the *wash/fill* position, move the probe from terminal 1 to terminal 2. Voltage should be present at this terminal. Voltage should appear at terminal 3 when the machine is filled with water, at which time there should be no voltage at terminal 2.

Resistance measurements must be made with power removed. Attach one probe of an $R \times 1$ ohmmeter to terminal 1 of the water-level control. If the tub is only partially filled, you must determine if the machine stopped during the fill or drain function. If the machine stopped during the fill operation, there should be zero resistance between terminals 1 and 2 and infinite resistance between 1 and 3. If it stopped during the pumpout phase it will be difficult to determine exactly which way the switch is made, to *fill* position (1 and 3 closed) or the *empty* position (1 and 2 closed). In this case, drain the water from the machine and start the test when the machine enters the wash/fill mode. Here terminals 1 and 2 are made (zero resistance) when the tub is filled with water. (This can be done with a hose to prevent applying power to the machine.) Terminals 1 and 2 should be open and terminals 1 and 3 should be closed.

Lid Switch

Most machines on the market will employ one of two types of lid switches. One popular type is a mercury switch where a small drop of mercury makes electrical contact with two wires inside a small glass ball. When the lid is raised, the lid switch is moved to a position that prevents the mercury from touching one or both wires. This opens the electrical circuit.

Another method is to trigger a small microswitch with a plunger on the lid. When the lid is closed, the plunger strikes the switch causing it to close. When the lid is open, spring tension opens the microswitch.

To test these switches, attach one probe of a voltmeter to terminal 1 of the water inlet valve (or any other point on the common side of the line). Attach the other probe to terminal 1 of the water-level control. With the lid closed, set the timer to the *wash/fill* position and apply power. You should read 120V on the meter. When the lid is opened, the meter should read zero.

To make resistance checks, remove all power and clip one meter lead to terminal 1 of the switch and the other to terminal 2. Raising the lid should result in infinite resistance. Closing the lid should cause the meter to show zero resistance.

Timers

Testing the timer appears to be the most difficult of all tests for the inexperienced technician. However, if you are careful and have a copy of the timing chart and wiring diagram, the tests are not so difficult. First of all, keep in mind that you should have some valid reason for testing the timer. Failure of a definite section of the timer (i.e., the machine fails to do some specified operation at the proper time) is solid justification for testing.

Voltage tests at sections in the timer that control the inoperative device should reveal whether that section is applying voltage as it should. Select a point where you can attach one meter probe to the common side of the line. With the other probe, touch the terminals of the timer that should apply power to the malfunctioning section. Move the timer through its cycle by hand. Resistance checks can be made by removing power and checking for continuity through the contacts of the timer during each portion of the cycle.

Motor Testing

Motor testing was described in detail in an earlier chapter of this book and will not be repeated here. Should you suspect motor failure, be sure to take the time to examine motor mountings and belts, as these can sometimes cause motor-shaft binding.

TROUBLESHOOTING

Electrical system troubles are located by observing the machine operation and comparing the actual operation with that specified in the timing sequence and schematic. Once you have become familiar with several models this process will be easy, and in many cases you may find you don't even need the timing diagram. The fact that the machine fails to perform a certain operation, or fails to sequence from one operation to another, will usually narrow the possible causes of failure. Electrical failures are usually easy to locate if you observe the machine operation closely.

Observing a complete wash cycle would take a considerable amount of time. This is done by most technicians only after other methods of troubleshooting have failed. A more popular method is to *short cycle* the machine. To do this, select the lowest possible water level and set the timer to the *wash/fill* position of the normal cycle. Start the machine and allow it to fill to the selected level. While the machine is filling, vary the water-temperature control and make sure that changing the control causes a corresponding change in water temperature. This is also a good time to discuss the problem with your customer. Once the machine has filled note that it automatically starts the wash/agitation cycle. If the water-level control is variable, turn the machine off and reset the water-level control to a higher setting. Restart the machine. It should once again start to fill. Again, the machine should stop filling and begin to agitate when the proper water level is reached.

If all is okay the timer can be advanced by hand to the end of the wash/agitate cycle, and the pumping and spinning operations can be observed. The machine is then allowed to fill for the rinse cycle and temperature check, as was done with the wash water. After this cycle has been determined to be operating correctly, the timer can be advanced to the spin phase. Observation time for the complete cycle of the electrical system is reduced to about one-third the time required for the machine to complete the wash cycle on its own.

Electrical problems will usually show themselves as a failure of the machine to complete the fill, agitate, pump, or spin operations. When a failure occurs, stop the machine and investigate further.

Chapter 7

Troubleshooting Automatic Washers

Methods of servicing the three major systems and their components were discussed in previous chapters. Isolating the problem to a sufficiently small area is sometimes a very difficult task for the inexperienced technician. Troubleshooting charts, especially those that list an exact part failure for a particular symptom, do not always lead you in the right direction. Because of differences from make to make and model to model, troubleshooting charts are somewhat limited in scope. They are useful, however, to the extent that they give the troubleshooter ideas.

Included in this chapter is a series of troubleshooting charts designed to be useful with any automatic washer. You have seen from earlier discussions that all automatic washers perform the same functions and, in general, go through the same cycles of operation. The specifics of the job may vary, but the job itself is the same. For example, all machines have a wash/fill period, a wash/agitate period, and a rinse/agitate period. To broaden the application of the troubleshooting charts shown in this chapter, we have written them to guide the technician in analytical checks of machine operation

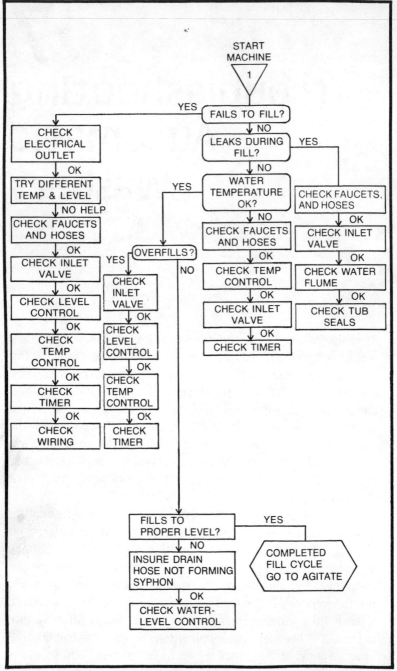

Fig. 7-1. Troubleshooting chart for the wash/fill cycle.

rather than point him to a definite component failure. Manufacturer's service literature, especially timing charts and wiring schematics, should be used to locate the mechanism that accomplishes the function listed in these diagrams. Specific checks of some components will be recommended where they are common to many washers.

The troubleshooting charts are most useful when the technician observes the machine in operation and notes discrepancies. To do this, the technician attempts to operate the machine in a normal manner investigating malfunctions as soon as they are noticed.

WASH/FILL

All machines must go through a period in which the machine fills with water prior to washing the clothes. The purpose of the wash cycle is not only to fill the tub with water, but to insure the water is at the correct level and temperature. If water level and temperature selection is permitted on the machine you are servicing, set the water-level control to low and water-temperature control to cold. Rotate the timer dial to the start of the normal wash cycle and start the machine. The trouble chart in Fig. 7-1 can be used as a checklist as you observe this operation.

Does the machine make any attempt to start the fill operation? Is there any indication that the machine is trying to start? If not, make sure the machine is connected to an electrical outlet and voltage is present at that outlet. If power is applied to the machine, vary the water-temperature control. Select hot water and notice if water enters the tub. Do the same with the water-level control. If varying either of these controls causes the machine to start filling, investigate the circuits associated with that control and the water inlet valve.

Should the controls have no effect, move the timer slightly. Sometimes the markings on the timer dial are not very accurate and you may be slightly off the *fill* position. If this does not reveal any clues, check hoses and faucets. Make sure they are properly connected and turned on. The next step is to check the water valve, as explained in an earlier chapter.

After this, check the water-level control and water. As a last resort, check the timer circuits. All these checks should be made with the aid of a schematic and a VOM.

Once the machine starts to fill, check for leaks by looking under and behind. If you see any water, check hoses and seals. To perform this check, it may be necessary to move the machine away from the wall and remove the back panel. With water entering the tub and no visible leaks, vary the setting of the water-temperature selector and see if changing the water-temperature control brings about a corresponding change in water temperature. Should temperature problems be encountered, check for proper connection to the water supply, the temperature control circuits (including the inlet valve), and the timer, in that order.

If everything is operating satisfactorily at this point, the only remaining check is the water-level control. To check this control allow the machine to fill to the lowest setting. When the water reaches the prescribed level and is turned off, move the control to a higher setting. Check for water shutoff at several settings of the water-level control. Should the water fail to shut off and the machine overfills, unplug the machine. If this fails to stop the flow of water, turn off the water valve behind the machine. When unplugging the machine does not stop the water from entering, you will usually find a stuck inlet valve. If unplugging stops the waterflow, look for problems in the water-level control, water-temperature control, or timer circuits.

Once the machine fills with water to the correct depth and proper temperature, and turns off the incoming water, the wash/fill operation is complete and you are ready to observe the wash/agitate function.

WASH/AGITATE

The purpose of the wash/agitate operation is to flex the soiled clothing in the wash water. Agitation should be smooth and steady. In machines equipped with automatic dispensers, detergents are released into the water during this period. If the machine features a recirculating filter, lint filtering takes place throughout the wash/agitate operation. (See Fig. 7-2 for troubleshooting this cycle.)

The machine should start the wash/agitate operation immediately after filling with water. Should this fail to happen, the most likely suspect is the water-level control. This control signals the mechanical system to start the agitate period. During the wash/agitate operation, the water-level control should be in the *full* position. This control can be checked by voltage and resistance measurements as explained in an earlier chapter.

If the water-level control is functioning as it should, your next move will depend on what type of machine you are servicing. Some machines use a solenoid to shift the mechanical components to the agitate mode. Others use the timer or motor-switching circuits. A wiring diagram should be used to trace the agitate circuit and make necessary repairs.

When the machine shifts into the agitate mode but operation is noisy, check for worn belts or frozen pulleys in the mechanical system. Bearings and internal transmission parts can also cause noisy operation. Pinpointing these defects usually requires major disassembly.

Troubles in automatic dispensing mechanisms can usually be traced to that device or to the timer. Use a wiring diagram and follow the circuit for these components. Make sure any tubing used to route the laundry agent from the tank to the tub is not clogged.

Most machines have some sort of brake to hold the spin basket steady during agitation. The spin basket tends to rotate if not held by the brake. This will reduce the effectiveness of agitation. Should this failure occur in a machine you are servicing, check the solenoids associated with the brake mechanism. Look for trouble in the brake linings or springs if the solenoids are working.

If the machine has operated properly up to this point, look for leaks around tub seals and hoses. This would complete the checkout of the agitate operation.

WASH/DRAIN

The wash/drain operation varies somewhat from model to model. In some washers this function starts immediately upon

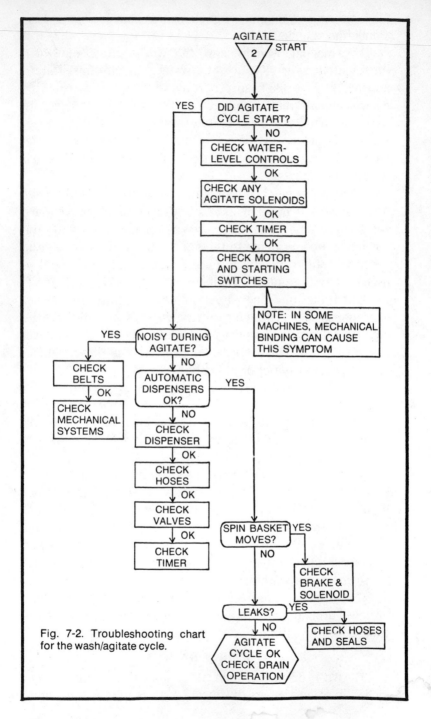

Fig. 7-2. Troubleshooting chart for the wash/agitate cycle.

completion of the wash/agitate cycle and continues until all water is removed. In other models, the water level recedes to about half capacity, then water is allowed to enter the tub to rinse soap scum. After this half-fill operation, all water is pumped from the machine. Another method pumps all water from the machine and refills it completely, after which a second pumpout is followed by the wash/spin period.

The wash/drain troubleshooting cycle begins with the completion of the wash/agitate period (Fig. 7-3). First note the action that occurs when the timer shows the wash/agitate time has ended. Compare this and subsequent actions to those shown in the timing diagram. If the machine fails to make any attempt to start pumping, check the drainhose for kinks or plugging and the lid switch for proper operation. Should it appear that the hoses are properly installed, check drive belts and the pump to make sure they are free to turn. The brake solenoid (on some models) and the water-level control can also cause the pump to fail. Some times you will find failures in the motor reversing circuits, motor-start switch, and the motor itself as the cause of pumping troubles.

Should the wash/drain operation start properly as outlined above, check tub seals and hose fittings for leaks while the water is draining. Completion of the drain period will be followed by a spin operation to remove the sudsy water from the clothing. If the machine starts the drain period, but fails to finish and enter the spin mode, check the lid switch, hoses, and water-level control, in that order.

WASH/SPIN

Following the pumping out of the wash water, the spinning operation starts to remove water remaining in the clothes. Most machines employ a spray/spin mode in which the clothing is revolved at high-speed while being sprayed with cool water. As shown in Fig. 7-4, the first things to check at this time are the starting of the operation, the amount of torque applied to the basket, operation of the lid switch, and the spraying (if used).

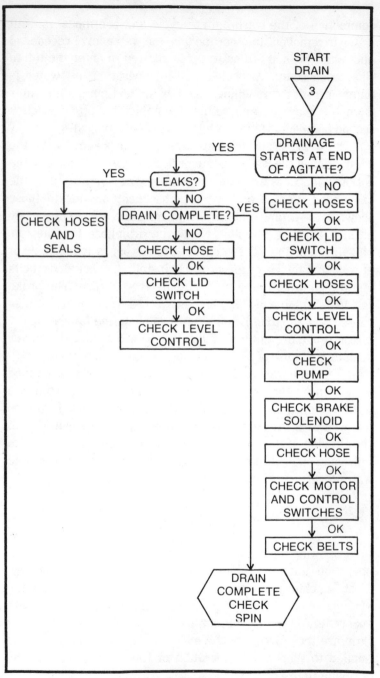

Fig. 7-3. Troubleshooting chart for the wash/drain cycle.

174

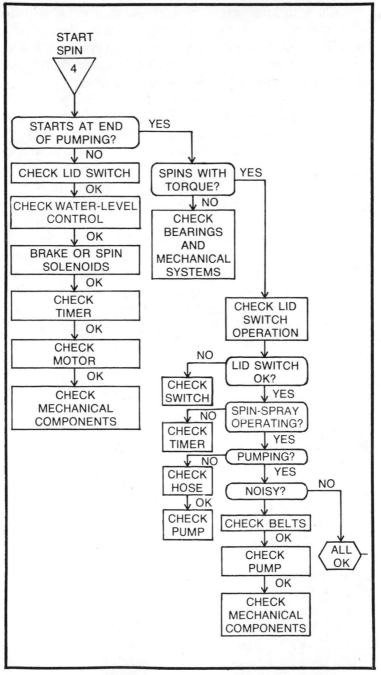

Fig. 7-4. Troubleshooting chart for the wash/spin cycle.

Failure to spin points to a check of the lid switch, water-level control, timer, and motor circuits. Mechanical failures in the transmission gearing or clutch system can also cause these symptoms. Check the solenoid-operated brake or spin solenoid on machines using these components.

As spinning starts, notice the amount of torque applied. Revolutions should build fast with an empty tub. Notice the speed with which revolutions are added and the smoothness of the spin. Faulty bearings in the mechanical components of the spin mechanism can cause sluggish and noisy operation. When the basket is spinning at its normal rate, check the lid switch by opening and closing the lid. The spinning should stop with the lid open.

On machines using a spin/spray operation, note if spraying is taking place. Although lack of spray can be caused by a faulty water-level switch or water inlet valve (on some machines), the timer is a more likely suspect. The balance of the machine affects its stability. An unbalanced machine tends to "walk." Make sure the load is balanced and check machine leveling. If this doesn't locate the problem, look for loose mechanical parts.

Noisy operation can be caused by belts or mechanical components. Observation is the best way to locate these faults.

At the end of the spin period the machine will usually go through a short pause (about one minute) after which the rinse and fill operations are started.

RINSE

Troubleshooting the rinse, fill, agitate, and drain operations generally follow the same procedures used for the wash cycle. Keep in mind here that any symptom seen in the rinse period should have been discovered in wash operation since in most cases the same components are used. When troubles are encountered in this period, and these same troubles are not duplicated in the wash mode, use a schematic diagram to locate components not common to both cycles. If an electrical component that would cause such a problem cannot be located, check the manufacturer's service literature for mechanical components that are not common.

SPIN

The final spin period is no different from the previous spin period, except that it lasts longer to remove the water more thoroughly. One manufacturer recommends a weight test to determine if the proper amount of water is being removed. Weigh the clothes prior to loading and again after washing. The wet weight should be approximately twice the dry weight.

8

Gas
Dryers

Any modern clothes dryer, whether powered by gas or electricity, uses three types of action to dry clothing: heat—applied by either a gas or electric heater to evaporate water in the clothing; air—applied by an electric blower or fan; and tumbling—applied by an electric motor to aid in air circulation and prevent burning the clothes. Notice that both tumbling and air circulation is accomplished by electrical devices, even in gas dryers. Gas dryers are those that use gas as a source of heat, as opposed to those using electrical sources.

GAS BURNERS

The source of heat for a gas dryer is a gas burner designed to operate on natural, manufacturered, or liquid-petroleum (LP) gas. Early models required the operator to manually light the burner each time the dryer was used. This inconvenience was elminiated by using a pilot-light system to automatically light the burner when the dryer was turned on.

Pilot

The basic pilot light and safety valve combination is still found on many dryers. Its simplicity makes it ideal for–

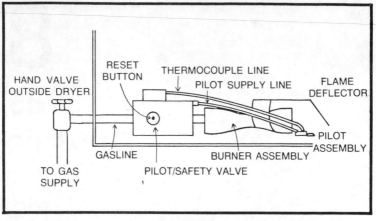

Fig. 8-1. A basic pilot and safety valve assembly.

starting our study of gas burners. Such a system is shown in Fig. 8-1. Most city plumbing codes require that each gas outlet be equipped with a manually operated shutoff valve outside the appliance. Gas enters through this valve and travels through piping to reach the pilot safety valve. The pilot safety valve has two outlets, one to the main burner and a smaller outlet to the pilot.

Both outlets are controlled by a thermocouple extending from the valve to the pilot flame. The job of the thermocouple is to sense whether the pilot is burning and shut off the gas whenever the pilot is out. A valve bypass is usually provided to manually light the pilot. This bypass is a pushbutton arrangement that allows gas to reach the pilot but not the burner. Pressing the button permits gas to flow to the pilot. Once lighted, the button must be held in until the thermocouple is hot enough to sustain the pilot. Once this happens the button can be released. With the pilot ignited, the burner will light any time gas is present.

A complete gas-burner control using this pilot system is shown in Fig. 8-2. Pilot operation is identical to that in the previous paragraph. A solenoid valve between the burner and pilot safety valve controls the electrical system. With the aid of the pilot and safety valve assembly, this valve allows the user to start the dryer from the control panel. The burner can then be completely hidden from view except when service panels are removed. Notice that the pilot gasline bypasses the solenoid control valve. The pilot remains lit when the dryer controls are in the *off* position.

With the pilot lit the operator can load the dryer and set the timer to the *start* position. The drying process will begin. The timer sends an electrical signal to the burner control valve causing the solenoid to open the line to the burner. The pilot ignites the fuel as it reaches the burner, and the dryer is in operation.

This pilot system uses a pilot with a constant flame. That means the pilot is consuming fuel even when the dryer is not in use for long periods of time. To cut down on the amount of wasted fuel, a variation of this system was designed that uses a pilot with a variable flame. When the dryer is not in use the pilot burns at a low level. The flame doesn't reach the thermocouple and, therefore, doesn't open the safety valve to allow gas to the burner. This condition is shown in Fig. 8-3. When the operator turns on the dryer, the control system allows more gas to reach the pilot and the flame level increases. With the pilot flame now reaching the thermocouple the safety valve will soon open and allow the burner to light.

This system has several disadvantages when used in automatic dryers and was not in production for very long. It is still used successfully with gas ovens, however.

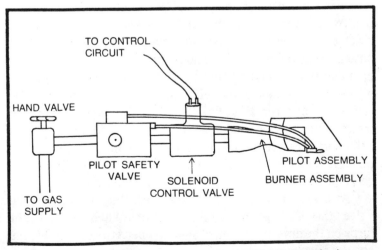

Fig. 8-2. A complete burner system using a solenoid control valve.

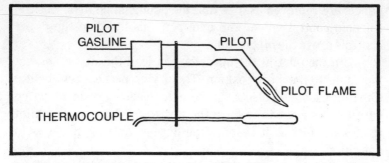

Fig. 8-3. A two-level pilot with a constant flame. With a low-level flame, the thermocouple is only warm, so no gas is supplied to the burner. With a high-level flame, the thermocouple is heated, turning on the safety valve and lighting the burner.

Combination Valve

One of the most popular burner control assemblies used with gas dryers is shown in Fig. 8-4. You will see this type of valve often if you service many automatic dryers. It has been in use for many years and is still found on some late-model machines. Although they may differ somewhat in size or shape, most combination valves work in the same manner.

Gas from the supply line enters the valve through the lower part of the regulator chamber. The purpose of the regulator is to continuously supply gas at a constant pressure. This valve controls the amount of pressure that reaches the pilot valve in the upper chamber of the regulator stage.

To light the pilot on this unit it is necessary to push the reset button and hold a match to the pilot burner. The reset button opens the pilot valve and allows gas to enter the lower chamber of that section. The supply line for the pilot is tapped off the lower section of the pilot valve chamber. Note that the solenoid valve remains closed and seals the main burner outlet. When the thermocouple has been heated by the pilot it will exert pressure on the pilot-switch diaphragm, engage the pilot latching cap, and hold the pilot valve open. The pilot will continue to burn when the reset button is released.

With gas pressure in the upper chamber of the solenoid valve, the dryer is ready for operation. When the operator turns on the dryer, the control system opens the solenoid valve and the pilot lights the burner.

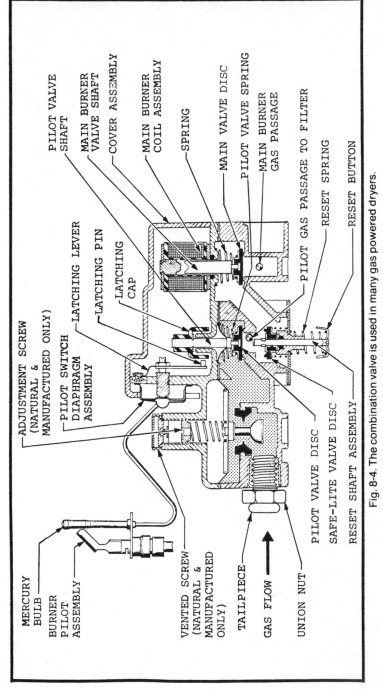

PILOT VALVE SHAFT

MAIN BURNER VALVE SHAFT

COVER ASSEMBLY

MAIN BURNER COIL ASSEMBLY

SPRING

MAIN VALVE DISC

PILOT VALVE SPRING

MAIN BURNER GAS PASSAGE

PILOT GAS PASSAGE TO FILTER

RESET SPRING

RESET BUTTON

ADJUSTMENT SCREW (NATURAL & MANUFACTURED ONLY)

PILOT SWITCH DIAPHRAGM ASSEMBLY

LATCHING LEVER

LATCHING PIN

LATCHING CAP

MERCURY BULB

BURNER PILOT ASSEMBLY

VENTED SCREW (NATURAL & MANUFACTURED ONLY)

TAILPIECE

GAS FLOW

UNION NUT

PILOT VALVE DISC

SAFE-LITE VALVE DISC

RESET SHAFT ASSEMBLY

Fig. 8-4. The combination valve is used in many gas powered dryers.

183

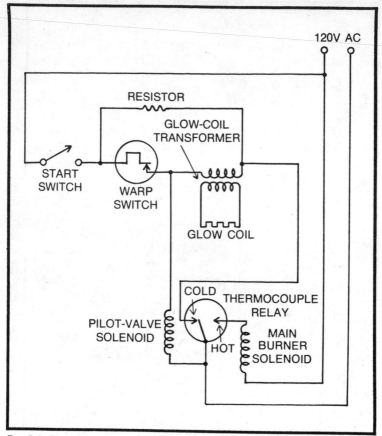

Fig. 8-5. The glow-coil ignitor consumes gas only when the dryer is in actual operation.

Glow-Coil Pilot

A standing pilot wastes fuel when the dryer is not in operation. To overcome this, several electrically operated pilot and burner-control mechanisms have been marketed. One type uses a hot coil of wire to ignite the burner. Another type uses electrical sparks. Both accomplish the same thing. No gas is consumed unless the dryer is in operation. Operation of a typical glow-coil control is explained with the aid of Fig. 8-5.

This system uses a pilot ignited by a wire heated with electric current. In Fig. 8-5 the pilot-control relay is actuated by the action of the thermocouple. When the dryer is idle the thermocouple relay is in the *cold* position. When the start switch is closed, current flows through the start switch, the warp switch (a thermostatic normally closed switch), and the pilot valve solenoid to the opposite side of the line. A second path exists through the resistor to the *cold* side of the thermocouple relay, through the cold contact, to the opposite side of the line.

Current in the primary of the transformer induces current in the secondary, and the glow coil starts to heat. The transformer is used to step down the 120V AC to a lower level, usually above 24V for safety. During this time, current through the pilot-valve solenoid causes that valve to open and supply gas to the pilot light. Heat in the glow coil builds rapidly and the pilot should light very quickly. When it does, the thermocouple relay senses the heat of the pilot flame and causes the center arm of the thermocouple relay to shift to the *hot* position. Current now flows in the main burner solenoid, opening that valve and supplying gas to the main burner, where it is ignited by the pilot.

The warp switch and resistor furnish a safety precaution for the burner control mechanism. Should the pilot fail to light within a short time, current through the resistor generates sufficient heat to cause the contacts of the warp switch to "warp" or bend open. This removes current from the glow-coil transformer and the pilot-valve solenoid, insuring that the gas flow to the pilot is stopped. The thermocouple relay would stay in the *cold* position, and heat from the resistor would hold the warp switch open. If this happens the dryer must be turned off for about 10 minutes. This will allow the warp switch to cool before you attempt to relight the pilot.

Notice that the design of this system is built around the safety of the pilot. Both pilot and main-burner solenoids must be energized for the dryer to operate. Should the pilot fail to light, all power to the burner controls is removed. Glow-coil ignition systems not only offer this safety feature but save fuel as well since there is no standing pilot.

Spark-Ignition System

Another electrically controlled pilot uses an electric spark to light the burner, thus eliminating the pilot light completely.

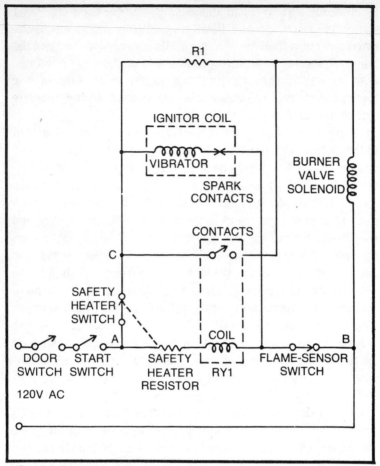

Fig. 8-6. The spark-ignitor system eliminates the pilot light completely.

A schematic of the spark-ignition system is shown in Fig. 8-6. This system is typical of those using a burner directly ignited by electrical controls.

Assume the dryer has been standing idle and the electrical components are in the position shown in Fig. 8-6. The machine is loaded, the door closed, and the start switch pressed. This allows current to flow through the two switches to point A—the junction of several parallel circuits. One circuit is from point A through the safety-heater resistor, the coil of relay RY1, and the flame-sensor switch to point B. Current through RY1 causes the contacts of the relay to close.

Current also flows from point A through the contacts of the safety-heater switch to point C. Again the circuit parallels. One path goes through the ignitor coil and the flame-sensor switch to point B. The vibrator is a small motor that opens and closes the spark contacts to produce an electric spark. The spark contacts are so positioned that the resulting sparks will ignite any gas present at the burner.

Another circuit exists from point C through the contacts of RY1 to the burner valve solenoid. This is paralleled by a path through R1 to the burner valve solenoid. Current through the coil of RY1 causes the contacts of RY1 to close, shorting R1. This produces heavy current through the burner valve solenoid, opening the valve and supplying gas to the burner.

Burner ignition should take place immediately. Heat from the burner is detected by a heat sensor that controls the flame-sensor switch. The flame-sensor switch opens as soon as the burner ignites. This removes current from the safety-heater resistor and the ignitor coil. However, current continues to flow through the contacts of the safety-heater switch. With no current through the coil of RY1, the contacts are open and current flows through R1 and the burner valve solenoid. The resistor (R1) reduces the current that passed through the coil during the ignition phase. The burner valve is designed so that less current is required to hold the valve open than is required for initial opening. The purpose of this design will be explained shortly. First let's see what happens if the burner fails to ignite.

Should ignition not occur quickly after the start switch is closed, the safety-heater resistor causes the contacts of the safety-heater switch to open. This removes all current from the ignitor coil and burner valve solenoid. This shuts off the gas and holds it off until the safety-heater resistor has time to cool.

Suppose the operator opens the door for a moment after ignition has occurred. Opening the door removes power from the burner valve solenoid. Gas to the burner is cut off. If the door is closed quickly enough, the only path for solenoid current is through R1 since the flame-sensor switch is still open. Since R1 doesn't allow enough current to open the burner valve solenoid, the system will not operate until the safety probe cools and closes the flame-sensor switch.

SERVICING AUTOMATIC BURNERS

Gas burners used in modern clothes dryers are reliable and can function for a long time without repairs. There are a few service steps that should be performed occasionally and, like any other electrical or mechanical device, repairs are sometimes needed. Cleaning and adjusting the airflow are the maintenance actions needed most often.

Over a period of time lint accumulates inside the dryer cabinet. Lint cannot only foul mechanical parts but it is also extremely flammable. So much so, in fact, that it is possible for large accumulations of lint to "flash" similar to gasoline vapors. This flash will usually occur during burner ignition and is sometimes followed by a less intense fire that burns and smolders in the lint. To prevent this, you should have a small vacuum cleaner in your toolkit and thoroughly clean all lint from inside the cabinet of every machine you service. This will not only help prevent fires but will reduce the need for servicing such failures as stuck centrifugal switches.

Adjusting the air supply to the burner is an easy service procedure. It will improve the efficiency of the dryer and reduce soot formation in the burner. Adjustment of air to the burner is accomplished as it is in other gas appliances. As shown in Fig. 8-7, gas enters the burner through a device that mixes the gas vapor with air. The flame should burn blue. Yellow flames indicate a dirty burner or an incorrect air setting. The air setting may be adjusted by loosening the setscrew and moving the damper to adjust the air inlet for clear, mostly-blue flames. Don't forget to tighten the setscrew.

Repairs to automatic burners depend on the type of pilot system used.

Standing Pilot

Failures, and likewise repairs, to the standing pilot are limited. In addition to cleaning and adjusting, failures of the pilot thermocouple are common. Thermocouple failure is usually indicated by the pilot light not staying lit when the reset button is released. Complete failure of the pilot indicates a gas flow stoppage. Look for closed valves or plugged lines. If the pilot remains lit but the burner fails to ignite, check the main burner valve.

Glow Coil

Due to its complexity, the glow-coil ignition system is subject to more failures than the standing pilot systems. Failures can occur in either electrical or gas components. Checks for opens and shorts will reveal most electrical problems. Gas component failures are best located by observing the overall operation of the burner.

To troubleshoot burners rapidly, you can classify malfunctions into one of four groups: pilot does not light, glow coil does not light; pilot does not light, glow coil is hot; pilot light ignites but goes out; main burner fails to light with pilot burning.

Failure of the glow coil to light can be caused by a defective glow coil, defective transformer, defective warp switch, defective stop switch, or defective thermocouple relay. To check any of these components, remove all electrical power from the dryer. With an ohmmeter across the contacts of the start switch, press the start button (see Fig. 8-5). The meter should show zero resistance. Now check the contact resistance

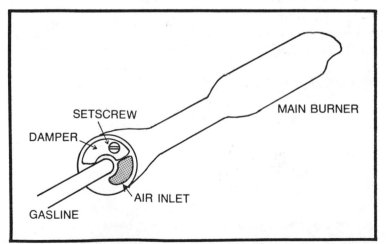

Fig. 8-7. Adjusting gas fires is accomplished by loosening the setscrew and moving the damper.

of the thermocouple relay in the same manner. Again the resistance should be zero. Once you have determined that these contacts are functioning properly, measure the resistance of the glow-coil (primary and secondary) transformer. If there are no opens or shorts check the wiring between the components. Although the design of glow-coil ignition systems vary from make to make, these checks should reveal any fault associated with a glow coil that will not heat. If you have made these checks correctly and still fail to pin down the fault, refer to the manufacturer's wiring diagram and check the electrical-control circuitry to the burner.

If the glow coil gets hot but the pilot fails to produce a flame, look for trouble in the area of the pilot burner. Check the continuity of the pilot solenoid. If the solenoid coil is good, restore power to the dryer and attempt to start the burner. Pay particular attention to whether you can hear the pilot valve operate. Should the pilot valve be operating properly, check for a clogged pilot line or gasline to the dryer. A valve may be closed.

Should the pilot ignite but fail to sustain a flame long enough to cause burner ignition, look for problems in the pilot thermocouple, the thermocouple relay, or the solenoid valve. This problem is sometimes caused by air drafts. Therefore, make sure all burner covers are in place and that no lint accumulates around the pilot.

Suppose the pilot functions as it should but the main burner fails to light? Refer once again to Fig. 8-5. Note the components in the main burner circuits. A defective thermocouple, thermocouple relay, or main burner solenoid could cause these problems. Check continuity of the relay contacts and the main burner solenoid.

Spark Ignition

Visual inspection, close observation of operation, and resistance checks are the surest ways to locate the faulty component. Keep in mind the operating sequence for the spark ignition described in conjunction with Fig. 8-6. Try to relate the failure to the function of one of the components in the figure. Divide the failure into one of three categories: burner fails to light, igniter fails to operate; burner fails to light, igniter operates correctly; burner lights but will not stay lit.

Complete failure of both the burner and igniter could be caused by troubles in either the electrical control system or the burner assembly itself. Before attempting to repair the burner assembly, make sure you have isolated the problem to that area. In a later discussion we will explain the troubleshooting procedures used in locating problems in the electrical control system. For now, assume we have already isolated the problem to the burner assembly and need to pinpoint the faulty component. Components that can cause a complete failure of the main burner and igniter include the flame-sensor switch, the safety-heater switch, the igniter coil and contacts, and RY1.

In cases of igniter failure always check the operation of the igniter coil and contacts. Usually, you can hear the igniter immediately after starting the dryer. Sparking of the contacts will make a destinctive clicking sound. Check for lint accumulated around the contacts and for burned or pitted contacts. The manufacturer's service literature will give the proper setting for the spark contacts.

If servicing the contacts themselves will not pinpoint the problem, resistance checks of associated components should be made. Check for continuity across the contacts of the flame-sensor switch and the safety-heater switch. If either switch is open during the startup period, power cannot reach the igniter. Resistance readings should also be taken across the igniter coil, the coil of RY1, and the safety-heater resistor.

Should these tests fail to locate the faulty component, use the wiring diagram for the machine and check the burner control circuitry.

The burner should ignite with the igniter operating properly. If not, look for problems in the burner valve solenoid, the contacts of RY1, and R1. Continuity checks across the relay contacts and resistance measurements of R1 and the solenoid coil should reveal the problem.

Should the burner light fail to remain lit, look for trouble in and around the burner as well as RY1, R1, and the solenoid coil. Visual inspection of the burner should locate any problems there. Resistance checks should pinpoint other problem areas.

GAS DRYER SYSTEMS

The gas dryer uses electrical power to operate the drive motor and control the sequence of events that occur during the drying cycle. Many of the components, such as timers and solenoids, are similar to those used in automatic washers. You will find, however, that timers and associated components used in dryers are usually simpler than those used in automatic washers. You will also encounter a new device—the thermostat. This is a very simple device used to control electric current. It is made of heat-sensitive bimetal that will open or close an electric circuit when heated, due to bending of the bimetal. In other words, it is a heat-sensitive switch.

Timer

Timers used with automatic dryers are built very much like those used in washers except that dryer timers have fewer devices to control and, therefore, have fewer contacts. For example, it is possible for an automatic dryer to be controlled by a timer with only three contacts. Figure 8-8 illustrates the wiring diagram of a timer.

Power from the 120V AC source is received by the timer through the door switch. Cam A opens and closes the contacts to the timer motor. This motor drives the timer shaft, causing the other contacts to open and close at the proper time. Cam A is closed whenever the timer dial is positioned at any point other than *off*. Cams B and C are designed to close the burner contacts after the motor contacts have been closed. Cam B supplies motor power and cam C furnishes power to the burner controls.

To operate a dryer with this type of control, the owner loads the machine, closes the door, and sets the timer to the desired drying time. The timer motor is energized and starts to drive the timer shaft. At the end of the drying time, the timer opens the contacts of cam A. This removes power from the motor and burner circuitry and completes the drying cycle.

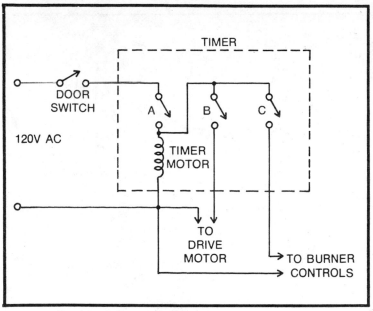

Fig. 8-8. A simplified wiring diagram for a dryer timer.

Timed Drying Cycle

The drying cycle of a typical dryer is shown in Fig. 8-9. This graph illustrates the temperature inside the drum at various times in a drying cycle of approximately one hour. Principles explained here apply to both gas and electric dryers. To understand this chart, assume the dryer has been off for some time and the ambient temperature inside the drum is 80°. At time *zero* the dryer is loaded with a bundle of wet clothes. This causes a drop in the drum temperature, as shown on the graph.

With the door closed and the machine started, the temperature inside the drum starts to rise. At about 120°, moisture in the clothes starts to evaporate as hot air is circulated. The burner (in gas dryers) continues to supply a constant heat at a much higher level than 120°. However, evaporation of moisture in the clothing causes a cooling effect. This phenomenon will result in the drum temperature "flattening out" at some level and remaining at that level until most of the moisture has been evaporated. The temperature remains fairly constant during most of the drying cycle.

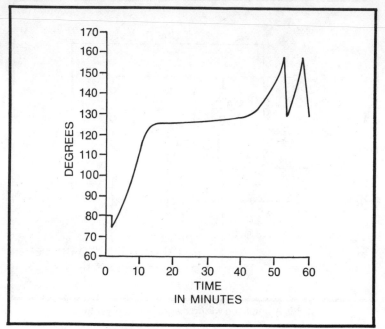

Fig. 8-9. The timed drying cycle.

Near the end of the cycle, when most of the moisture has been removed, the temperature starts to rise quite rapidly. This is due to the lack of evaporation. At some point (usually around 160°) the temperature reaches a level that causes the contacts of a bimetal thermostat to open. This thermostat is in the circuit to the heating device. Power is removed from the burner solenoid in gas dryers. With the burner off the temperature inside the drum begins to drop. If the clothes were removed at this time they would still be wet.

Although the burner was extinguished, the blower and drum continue to function. They tumble the clothes and circulate the air. As the temperature continues to drop, the thermostat cools and the contacts close. The burner is relit and the temperature rises to open the thermostat. This process continues until the timer cycles to the *off* position or the clothes are removed from the dryer.

Most fabrics contain moisture amounting to about 5% of their weight. Ideally, the timer should be set to a time that would dry the clothing to that level of moisture. This also means that the clothes in any one load should be about the same weight. If not, some will overdry and some may underdry.

The drying cycle described here is known as a *timed cycle*. A timed cycle is one that runs for a selected time and stops. The drying process is dependent on the time the load has been in the dryer rather than the actual moisture content of the clothing.

Automatic Drying Cycle/Auxiliary Thermostat

Many older dryers, although called automatic, feature only the timed-dry sequence. The development of synthetic fabrics led to a need for different drying cycles for different materials. This need could be overcome to some extent by selecting different drying times. This is not automatic in the true sense of the word, however. Several methods have been used that employ a variable drying time. The operator may select one of two or more drying cycles to conform to the type of fabric being dried.

Figure 8-10 shows the control circuitry used with one automatic dryer. This system makes use of two thermostats to

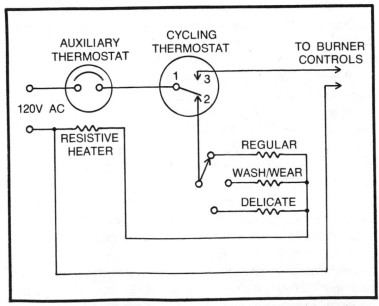

Fig. 8-10. An automatic drying control using an auxiliary thermostat.

control the time the burner is on. The cycling thermostat is actually a heat-sensitive single-pole, double-throw (SPDT) switch. Physically, this thermostat is located where it is exposed to air exhausted from the drum. Electrically, the contacts of the thermostat are wired so contacts 1 and 3 supply current to the burner controls while contacts 1 and 2 supply current to the resistor network. The cycling thermostat is designed to close contacts 1 and 3 and open contacts 1 and 2 at temperatures below 120°. At 130° contacts 1 and 3 open and contacts 1 and 2 close.

A second thermostat, sometimes called an auxiliary or bias thermostat, is also used in the control circuitry. This thermostat has its own heat supply in the form of a resistive heater. The contacts are located in the voltage supply line to the cycling thermostat. Current through the resistive heater controls the heat applied to the thermostat and, therefore, controls the opening and closing of the contacts. On this particular model, three different sizes of resistors can be alternately placed in series with the thermostat heater by means of switch contacts—one each for regular, wash-and-wear, and delicate fabrics. The smaller the series resistance, the more current allowed to flow through the resistive heater, causing it to heat faster. Current through the heater is controlled in such a manner that the contacts of the auxiliary thermostat open when the clothing is dry.

When the dryer is loaded the operator selects one of the three automatic cycles (depending on the type of clothing) and starts the machine. Contacts 1 and 3 of the cycling thermostat are closed as are the contacts of the auxiliary thermostat. The burner is fired—generating heat for the drying process. No current is flowing through the resistive heater since contacts 1 and 2 of the cycling thermostat are open.

When the exhaust temperature reaches 130°, contacts 1 and 2 of the cycling thermostat close and contacts 1 and 3 open. This shuts off the gas supply to the burner and current flows through the resistive heater. With the heat source removed, the exhaust temperature drops. The temperature of the auxiliary thermostat increases due to current through the resistive heater. At a point where the temperature of the drum

196

exhaust drops to 120°, the contacts of the cycling thermostat reverse—closing 1 and 3 and opening 1 and 2.

The burner is started and heat is applied to the drum. Since current has been removed from the resistive heater, heating of that unit ceases. The controls remain in this state until the drum exhaust reaches 130° and the contacts of the cycling thermostat are reversed—causing the auxiliary thermostat to heat. Between periods of current flow the auxiliary thermostat does not have time to dissipate very much heat. Therefore, each time current flows through the heater it is at a somewhat higher temperature than it was at the beginning of the previous period.

This cycle alternately applies heat to the drum and auxiliary thermostat. Each cycle is shorter due to residual heat in the drum. At some point, the auxiliary thermostat will reach a sufficient level to open its contacts. When this happens all power to the control circuit is removed and the drying process is ended. The resistance of the series resistors is determined by the manufacturer to insure proper drying time for selected fabrics.

Automatic Drying Cycle/Cycling Thermostat

A variation of the control circuit for an automatic drying cycle is shown in Fig. 8-11. Operation of these controls depend

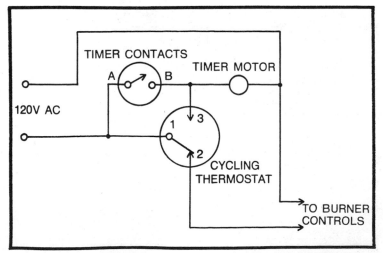

Fig. 8-11. An automatic drying control using a cycling thermostat.

on the switching of an SPDT thermostat. Upon initial loading, the machine is in a cooled state and contacts 1 and 2 of the cycling thermostat are closed. When the timer is placed to the automatic-dry cycle, timer contacts A and B are closed. This starts the timer. Current through contacts 1 and 2 allows the burner to start.

After about 15 minutes, timer contacts A and B open. Power is removed from the timer motor and the timer ceases to advance. The burner and main drive motor continue to heat

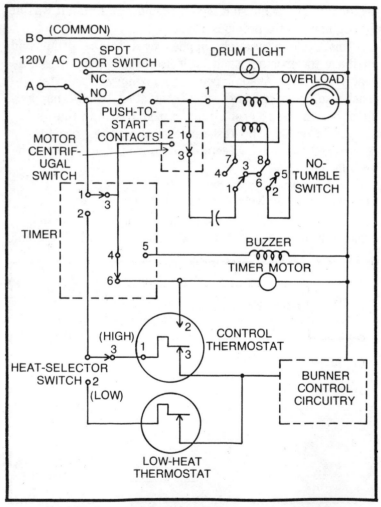

Fig. 8-12. A wiring diagram for a typical automatic gas dryer.

and dry the damp clothes. When the temperature starts to peak (as shown in Fig. 8-9), the cycling thermostat opens contacts 1 and 2 and closes contacts 1 and 3. Removing power from the burner extinguishes the flame. The motor circuit remains closed and tumbles the clothing, however. With contacts 1 and 3 of the thermostat closed, the timer motor is started and the timer starts to advance. This phase of operation (burner off, timer running, and main motor running) continues for about 5 or 10 minutes. This is to allow the clothes to cool while tumbling to prevent wrinkling. At the end of this period the timer will cut off all power and stop the cycle.

ELECTRICAL SYSTEM

A complete wiring diagram of a typical automatic gas dryer is shown in Fig. 8-12. This machine represents a "top of the line" model with several convenience features. It has two automatic cycles and one timed cycle. Four heat selections are available: low, high, no heat, and automatic. In the *no heat* mode the burner is not energized but the motor circuits are. This allows the clothes to be tumbled or fluffed without applying heat. Selection of either automatic cycle, one for regular fabrics and one for permanent-press fabrics, will automatically select the *high* mode. Another feature of this machine is called a *stop and dry* mode. In this operation, heat is applied but the clothes are not tumbled. Instead, a drying rack is placed in the drum and very delicate fabrics are placed on the rack. Operation in the stop and dry mode applies warm air to the load.

Timed Drying Cycle

In Fig. 8-12, 120V AC is applied to the dryer between points A and B. The *B* line will serve as a common throughout the machine. The SPDT door switch serves two functions. When the door is open, the normally closed (NC) contact is made and the drum light is on. At the same time, the normally open (NO) contact is open, removing power from the rest of the machine. With the door closed, power is available at contact 1 of the timer and at the push-to-start (momentary-contact) switch.

Notice how the timer contacts are used to sustain motor operation once it has been started. Whenever the timer dial is moved from the *off* position, contacts 1 and 3 close, supplying power to the centrifugal start switch at contact 2. This switch remains open, however, until the motor reaches its proper operating speed. Therefore, contacts 1 and 3 of the timer cannot furnish starting power. Built into this timer is a momentary-contact switch on the timer shaft. Once the timer has been moved from the *off* position, pushing the timer shaft closes the momentary-contact switch. Power is supplied through the momentary-contact switch and contacts 1 and 3 of the motor centrifugal switch. (The function of the no-tumble switch will be discussed later.) Once the motor reaches operating speed, contacts 1 and 3 of the motor centrifugal switch open and contacts 1 and 2 close. The momentary-contact switch can now be released and the motor will continue to run. Contacts 1 and 3 will remain closed until the timer motor drives the timer switch to the *off* position.

Contacts 1 and 2 of the timer close at the same time contacts 1 and 3 close. Contacts 1 and 2 will open about five minutes before the end of the drying cycle. From contact 3, a jumper supplies power to the switching arrangement at contacts 4, 5, and 6. During timed-dry operations, contacts 4 and 6 remain closed until the end of the drying period. When contacts 4 and 6 open, contacts 4 and 5 close and a buzzer notifies the operator that the drying cycle has been completed.

Heat selection during the timed drying cycle is achieved by the heat-selector switch. In the *low* position, contacts 1 and 2 are closed, routing power through the low-heat thermostat to a set of contacts on the motor centrifugal switch. *High* heat is selected when contacts 1 and 3 are closed. Current for the burner unit is then through the control thermostat, motor centrifugal switch, and overload. When *timed-dry* is selected, timer contacts 4 and 6 remain closed during the entire operation. Therefore, the timer motor continues to drive regardless of the position of the contacts on the control thermostat. In timed-dry operations using high heat, the control thermostat will cycle the burner on and off but has no effect on the operation of the timer.

Automatic Dry/Regular Fabrics

To dry regular fabrics using the automated features of this machine, the cycle for automatic dry/regular fabrics is selected. Contacts 1 and 2 of the timer, the push-to-start contacts, and the motor centrifugal switch function just as they did in the timed-dry cycle. Contacts 1 and 2 will close when the timer is positioned away from the *off* position. In this cycle these contacts are opened by the timer cams shortly before the cycle ends.

In automatic operation the heat-selector switch is set to the *high* position, eliminating the low-heat thermostat from the circuit and selecting the control thermostat. Burner-control current is supplied through contacts 1 and 2 of the timer, the high-heat contacts of the heat-selector switch (1 and 3), contacts 1 and 3 of the control thermostat, the motor centrifugal switch, and the overload. Notice that the overload is always in the circuit for burner-control current regardless of which cycle is selected. The overload limits the amount of heat generated by the burner, opening the circuit and removing power any time the temperature rises to a specified maximum. This prevents overheating and damage to the machine or clothing.

The most obvious difference between the timed-dry cycle and the automatic-dry cycle is the timing of the opening of contacts 4 and 6 in the timer motor circuit. When the regular-fabric automatic cycle is selected, these contacts close at the start of the cycle but open after a short period of operation. This removes current from the timer motor and stops the timer. The other operations of tumbling and heat application continue. When the temperature in the drum rises to a point indicating most of the moisture has been evaporated, the control on the cycling thermostat opens at contacts 1 and 3 and closes at contacts 1 and 2. As a result, the burner is extinguished and the timer motor starts while the main motor and blower continue to operate. Temperature in the drum will begin to drop after a short time and contacts 1 and 2 will open while 1 and 3 close.

At this time, the timer is stopped and heat is applied to the drum. When the drum temperature reaches the specified level, the control-thermostat contacts shift positions—the burner goes out and the timer comes on. This off-again, on-again operation continues until the timer advances to where the cam closes contacts 4 and 6 in the timer motor circuit. Operation from here to the end of the cycle is the same as it was for the timed-dry cycle.

Automatic Cycle/Delicate or Permanent-Press Fabrics

Operation of this cycle is similar to that used for automatic drying of regular fabrics. The major difference is the operation of contacts 4 and 6 in the timer motor circuit. In the regular fabric cycle, note the differences shown in Fig. 8-12. Since these contacts are opened at a point near the end of the cycle, during permanent-press cycles less heat is applied to the clothing during the last few minutes of drying. This reduces wrinkling by tumbling the clothing under low-heat conditions.

Fluff Cycle

This cycle allows the main motor and timer circuits to operate while disabling the burner-control circuits. The result is clothing tumbled in an air stream, but no heat is supplied to that air. Some housewives use this cycle to "finish dry" permanent-press fabrics or remove lint that clings to some synthetic fabrics when dried under high-heat conditions.

During this cycle, the contacts of the heat-selector switch remain open to prevent current from reaching the burner. Timer motor and drive motor circuits function the same as they do for timed-dry operations.

No-Tumble Dry

Some manufacturers have designed machines to meet the needs of those who prefer a drying operation that applies heat to the load without tumble action. This cycle is useful for items such as knit sweaters that might stretch out of shape in a tumble dry cycle.

To accomplish this, a special clutch is used on one end of a double-shafted motor. Figure 8-13 illustrates how this motor is connected to the drum and fan. One end of the shaft is equipped with a standard V-belt pulley attached to the shaft by a setscrew. The V-belt is used to drive the fan any time the motor is running. The opposite motor shaft is equipped with a special ratchet-type clutch that allows it to freewheel when the motor is turning in the reverse direction.

For example, when the motor is turning clockwise, both pulleys are locked to the motor shaft and both the fan and drum are driven. If the motor turns counterclockwise, the fan will still turn. However, inside the drum pulley is a ratchet mechanism that allows the motor shaft to slip inside the drum pulley. When the motor is turning in this direction, the drum pulley does not turn and the drum remains stationary. In using this mode, a rack is placed inside the drum and the article to be dried is placed on the rack. Other dryer controls operate as they would for any other cycle.

Figure 8-12 illustrates the wiring necessary to accomplish this type of operation. In the start circuit for the main drive motor is a switch labeled NO TUMBLE. For normal dryer operation the DPDT switch is made from 1 to 3 and 2 to 5. This establishes the normal polarity of the start winding in relation to the run winding. It results in normal motor rotation. Placing the switch in the *no tumble* position connects contacts 1 and 4 and contacts 2 and 6. This reverses the polarity of the start switch, and the motor will now turn backwards. Since the clutch slips, the drum cannot turn.

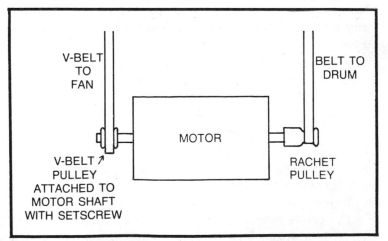

Fig. 8-13. A typical motor and pulley arrangement used for no-tumble operation.

Troubleshooting The Electrical System

Voltage and resistance measurements will reveal the source of most troubles that develop in the electrical system. When taking voltage measurements, remember that power is being applied to the circuits. Take precautions to avoid shocks. If resistance readings are to be made, disconnect all electrical power.

AIR CIRCULATION

In order to dry clothing properly without overheating, adequate air circulation is required. A fan capable of moving several cubic feet of air per minute is used to move air through an arrangement of vents and ducts. Although the details of the air supply system vary from dryer to dryer, the overall operation is the same. Room air is drawn into the machine, heated, passed over the clothing, and expelled. Figure 8-14 is typical of these systems.

A large fan or blower is placed in the exhaust vent and pushes air out of the machine through the vent ducts. A low-pressure area is created in front of the blower. The draft of this low-pressure area pulls air from the room into the cabinet and through the heater. The tendency of hot air to rise and the action of the blower force the hot air up through the interior venting of the machine and through the vent holes in the drum. Once inside the drum, hot air picks up moisture and lint before being drawn into the lint trap, into the blower, and out of the machine.

Although simple in design, the lint trap plays an important part in the drying process. Lint picked up in the drum is very flammable. If not removed, it could collect in the blower and exhaust vent and result in a serious fire. Also, it can block the vent and prevent the passage of air, resulting in poor drying efficiency. Always check the condition of the lint trap when servicing automatic dryers. Instruct your customers on proper cleaning procedures. Most manufacturers recommend cleaning the lint filter after each load.

To operate properly, the air circulation system requires that all gaskets and seals be in good condition and all ducts be free of obstructions. Anything that restricts the airflow within

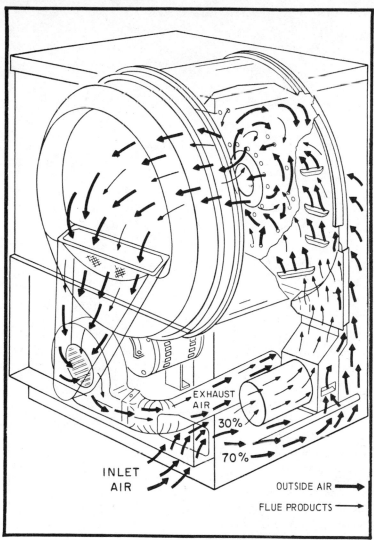

INLET
AIR

EXHAUST
AIR

30%

70%

OUTSIDE AIR ➡
FLUE PRODUCTS ➡

Fig. 8-14. A typical air circulation system used in gas dryers.

the machine will reduce drying efficiency. Likewise, a leaky gasket or seal can let air enter the system at some point other than the burner. This air, at room temperature, also reduces drying efficiency.

Chapter 9

Electric Dryers

The general operating characteristics of an electric dryer are not very different from those of a gas dryer. The biggest difference is in generating heat. You will probably find the electric dryer much easier to understand than the complex gas dryer. Although some electric dryers can operate on standard 120V house current, most (including those discussed here) operate on 240V. Electric dryers operating on 120V power are not as efficient as their 240V counterparts. The lower voltage causes more power to be used and requires a longer drying time. These disadvantages have decreased the popularity of these machines; therefore, we limit our discussion here to those machines using 240V power.

ELECTRIC BURNER

Heat generation in the electric dryer is accomplished with a high resistance wire-type heating element (Fig. 9-1). The wire is formed in a small coil (similar to a door spring) and can be strung in a large circle behind the dryer drum or formed into a unit that fits inside the dryer intake ducts. Operation is the same. Only the location is different.

The element may be a single or double strand between 3000−6000W. The double strand elements are used on

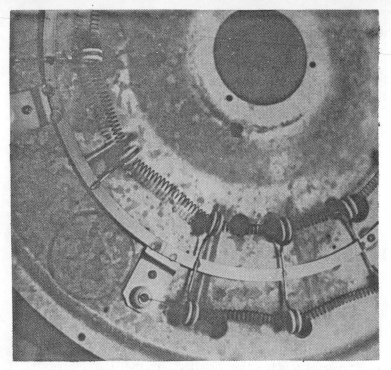
Fig. 9-1. A typical heating element used in electric dryers.

machines featuring two or three temperature settings where these settings are controlled by the amount of current drawn by the element. Temperature control can also be achieved by a thermostat.

Single Element/Timed Dry

The circuit for a single element dryer offering only a timed-dry cycle is shown in Fig. 9-2. Power is supplied at points L1 and L2 at 240V AC. The current to the element must flow through the timer contacts and safety thermostat before reaching the heating element. To prevent heat damage to the clothes, the safety thermostat will open the element circuit in case of overtemperature. The timer contacts stop the current flow at the end of the cycle. A centrifugal switch protects the clothing in case of motor failure. Without this device, should the dryer be turned on and the motor fail to start, the clothing and machine could be damaged by the high temperatures. The

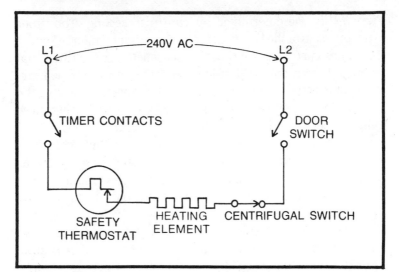

Fig. 9-2. A simplified wiring diagram for a single-element timed-dry operation.

door switch will open the current path whenever the dryer door is opened.

A variation of this circuit is shown in Fig. 9-3. This dryer uses two thermostats and a single heating element to obtain three cycles of operation, all of which are timed. No automatic control of the drying time is employed. Instead, each cycle

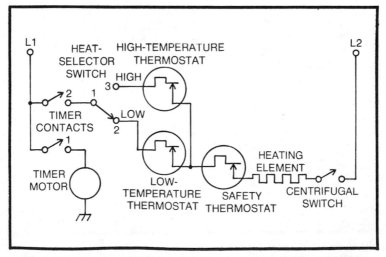

Fig. 9-3. Single-element, timed-dry operation using two thermostats.

runs for a time determined by the timer setting. The timer contacts control the timer motor and heating element. Contact 1 supplies current to the timer motor whenever the timer dial is moved from the *off* position. Contact 2 furnishes heating-element current. The cam for contact 2 is cut so that during the normal fabric's cycle the timer will open contact 2 about three minutes before contact 1. During the permanent-press cycle, contact 2 opens about five minutes before contact 1. In both cases, current to the element is interrupted prior to motor cutoff. This prevents "baked in" wrinkles. Tumbling for a short period in the warm air tends to shake out wrinkles left by washing and drying.

As mentioned previously, this is a three-cycle machine. The heat-selector switch and the two heat-control thermostats make this possible. When high heat is selected, contacts 1 and 3 of the heat-selector switch are closed; low heat closes 1 and 2. Contacts 2 and 3 are open during the *fluff dry* cycle, removing all current to the heater element. The high-temperature

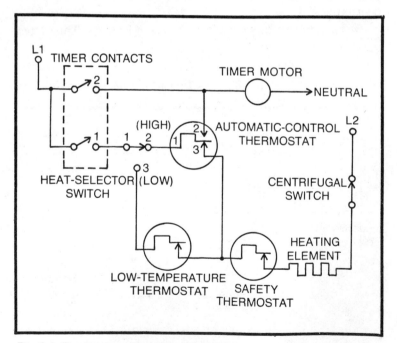

Fig. 9-4. The heating-element control circuit for a single-coil dryer with automatic control.

thermostat opens at 145° and closes at 135°. The low-temperature thermostat opens at 135° and closes at 125°. The safety thermostat and the centrifugal switch in this example operate in the same manner discussed in conjunction with Fig. 9-2.

Single Coil/Automatic Dry

Figure 9-4 shows the heating-element control circuit for a single-coil dryer with automatic control. In addition to the regular timed cycle, this circuit features two automatic drying cycles. One is for regular fabrics and the other is for permanent-press fabrics. The heat-selection options found in the previous control system are also found in this system.

The only difference in the two automatic cycles is the timing for opening timer-contact 2, which supplies power to the timer motor. Timer-contact 2 opens about three minutes before the end of the drying cycle for regular fabrics and about five minutes before the end of the drying cycle for permanent-press fabrics. This removes power from the heating circuit and allows time for the clothes to cool for easier handling.

The timer motor opens contact 2 about 15 minutes before the end of the cycle to achieve automatic operation. The automatic-control thermostat now controls both the heating element and the timer motor. Contacts 1 and 3 of the automatic-control thermostat close when the temperature falls below a preset level. Current for the heating element is now supplied through these contacts and the safety thermostat.

As the temperature rises, contacts 1 and 3 in the automatic-control thermostat open and contacts 1 and 2 close. This removes current from the heating element and supplies it to the timer motor. The timer advances and the drum temperature drops. When the temperature is low enough, the automatic-control thermostat is reset and the heating element raises the drum temperature. This sequence continues until the timer advances to the *stop* position and opens timer-contacts 1 and 2.

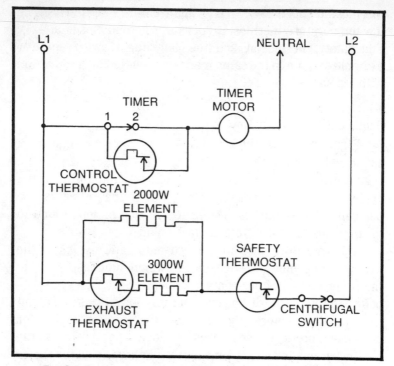

Fig. 9-5. The wiring diagram for the dual-element heating circuit.

Dual-Element Heater

Figure 9-5 is a wiring diagram for the dual-element heating circuit. This element offers a closer control of drum temperature by applying a more constant supply of heat.

In the timed cycle, timer-contacts 1 and 2 remain closed. This supplies current from L1 of the 240V supply to the timer motor. Current for the heating element is supplied between L1 and L2 and passes through the safety thermostat and the centrifugal switch.

To achieve automatic operation, timer-contacts 1 and 2 remain closed for the first 15 minutes of the drying cycle. When these contacts open, the heating element is controlled by the exhaust thermostat, and the timer is controlled by the control thermostat. These two thermostats open at about the same temperature. The control thermostat opens the timer motor circuit. The exhaust thermostat opens the circuit for the 3000W heating element. This circuit differs from the single-element circuit in that it continues to supply heat when the timer motor is energized. This heat passes through the 2000W heating element and is therefore somewhat lower. Other than this difference, the single-element and dual-element circuits operate the same.

ELECTRIC DRYER WIRING

Although there are many similarities, there are also major differences between gas-dryer wiring and electric-dryer wiring. The differences arise from the fact that electric dryers operate on a three-wire 240V line, whereas gas dryers operate on the conventional 120V line.

At this time you should review the chapter covering the service entrance and the 240V service. In the next chapter we will look closer at the 240V dryer hookup, but at this point we want to concentrate our attention on the circuits within the dryer itself. To do this, two dryers will be used as examples. One features only timed-dry cycles. The second has automatic dry and several other features as well.

Timed Dry

Electric power is furnished to the dryer in Fig. 9-6 through a three-wire hookup. Between lines L1 and L2, 240V AC is supplied. Between the neutral and either L1 or L2 will be 120V AC. Close examination of Fig. 9-6 will reveal that the drive motor and timer motor circuits are supplied with 120V since they are connected between L1 and neutral. The heating element circuit takes 240V, between L1 and L2.

This model offers two cycles with three heat selections. When the heat-selector switch is set at *low*, contacts 1 and 2 are closed and the low-temperature thermostat is in the circuit. If *high* is selected, contacts 1 and 3 of the heat-selector switch are closed and the high-temperature thermostat is in control. For *fluff dry*, both contacts are open and no heat is supplied. The safety thermostat protects against overheating in all cycles. The centrifugal switch prevents heat from being applied until the motor is up to its proper operating speed. Contact 3 of the timer is closed when the timer is moved from the *off* position.

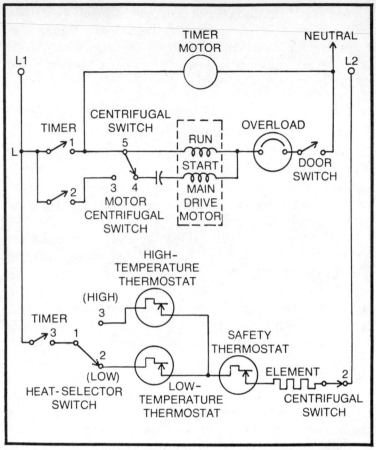

Fig. 9-6. Wiring diagram for a timed-dry machine.

It remains closed until about three minutes before the regular cycle is completed. If the permanent-press cycle is selected, timer-contact 3 opens about five minutes before motor cutoff.

Timer-contact 1 is a momentary-contact switch used to start the timer and main drive motors. Once the motors start, current is furnished through timer-contact 2 and contacts 3 to 5 of the motor's centrifugal switch. This circuit sustains current for the drive motor and timer motor throughout both cycles.

Automatic Dry

The wiring diagram of a dryer featuring two automatic drying cycles is shown in Fig. 9-7. One timed drying cycle and three heat levels are provided. This machine also has a *dry without tumble* feature. Essentially, this machine is the same as shown in Fig. 8-12, with two exceptions: the electric heating system, and the manner in which 120V AC for the control circuit is tapped from the 240V supply. Tracing the circuit in the diagram will illustrate these differences.

Between L1 and neutral, 120V AC is routed to the main drive motor and control circuits. Voltage is supplied from L1

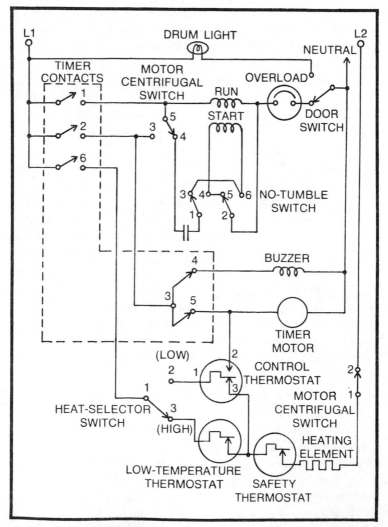

Fig. 9-7. A wiring diagram for a dryer featuring two automatic cycles.

directly to one side of the drum light. The other side of the drum light is connected to a contact on the SPDT door switch. When the door is open, this contact closes and current flows through the light. Closing the door removes power from the light and connects the motor circuit between L1 and neutral.

Two timer contacts are in the motor control circuit. Timer-contact 2 is closed when the timer dial is set to any position other than *off*. This places 120V (to neutral) at contact 3 of the motor centrifugal switch. Pressing timer-contact 1 (momentary contact) supplies the same voltage to contact 5 on the switch. With the motor stopped, voltage is supplied through contacts 4 and 5 of the motor centrifugal switch and the no-tumble switch in the start winding circuit. Once the motor reaches operating speed, contacts 4 and 5 of the motor centrifugal switch open and contacts 3 and 5 close. Voltage is then supplied through contact 2 of the timer and contacts 3 to 5 of the motor centrifugal switch.

From timer-contact 2, a jumper connects the voltage to timer-contact 3. Contacts 3, 4, and 5 control the buzzer and the timer motor. Contacts 3 and 4 are closed for a short period at the end of each cycle to operate the buzzer. Contacts 3 and 5 are closed whenever the timer dial is moved from the *off* position. During timed-dry operation these contacts remain closed until the end of the drying period. If either automatic cycle is selected, contacts 3 and 5 are closed for the first few minutes of the drying cycle. This allows the timer motor to run while the heating element raises the drum temperature to the drying level.

Timer-contact 6 furnishes current to the heating element through the heat-selector switch. In the *low* position, contacts 1 and 3 are closed, routing power through the low-heat thermostat, the safety thermostat, the heating element, and a set of contacts in the motor centrifugal switch. High heat is selected when contacts 1 and 2 are closed. Current to the heating element is supplied through the control thermostat. In timed-dry operations the thermostats turn the element on and off as necessary to maintain drying temperature. If *fluff dry* is selected, contacts 2 and 3 are open and the dryer runs in a timed cycle with no heat applied.

Opening and closing of timer-contact 6 is determined by the cycle of operation. This contact closes when the timer dial is moved from the *off* position and opens about five minutes before the end of the timed-dry or regular fabric cycle. It also opens about three minutes before the end of the permanent-press cycle.

Automatic operation depends on the cycling of the control (high-temperature) thermostat. When an automatic drying cycle is selected, timer-contacts 3 and 5 close at the start of the cycle but open after a short period. This removes current from the timer motor. The timer stops—tumbling and heat application continues. When the drum temperature trips the control thermostat, power is removed from the heating element and applied to the timer motor. As the timer motor runs, the drum temperature starts to fall. When a specified limit is reached, the control thermostat will close contacts 1 and 3 and open contacts 1 and 2. Heat is again applied while the timer stands still. This off again, on again operation continues until the timer advances to the *off* position.

The no-tumble feature is achieved with a reversing switch in the motor-start winding circuit just as it was with the gas dryer. Dryers with these cycles are equipped with a reversible motor and a special clutch allowing the motor to rotate backwards while the drum remains stationary. (For a complete discussion of that cycle, refer to the section in Chapter 8 dealing with no-tumble operation.)

SPECIAL FEATURES

Manufacturers have gone to great lengths in recent years to improve the automated features on their deluxe models. These unique features vary from model to model. They effect the overall operation so little that only the two most popular features are discussed here.

Antiwrinkle Cycle

At least two popular makes of automatic dryers feature a unique cycle that goes a long way in achieving a truly automatic wrinkle-free drying cycle. A special timer used in

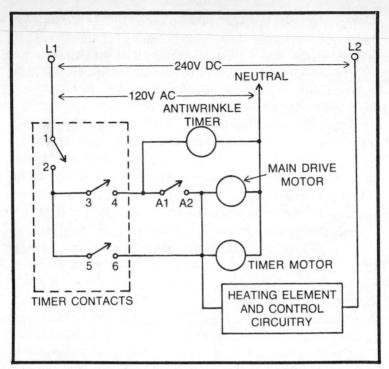

Fig. 9-8. A simplified diagram of an antiwrinkle control circuit.

these machines allows the load to be tumbled periodically for short durations at the end of the automatic dry cycle. The diagram of the control circuit for this feature is shown in Fig. 9-8.

Contacts 1 and 2 are the usual on/off contacts of the timer. Through these contacts current is furnished to the timer and main drive motor. The antiwrinkle feature adds contacts 3, 4, 5, and 6 on the regular timer and a second timer motor (contacts A1 and A2). When the timer is moved from the *off* position, contacts 1 and 2 close, completing the circuit to the controls. At the end of the automatic dry cycle, contacts 5 and 6 open and contacts 3 and 4 close. With contacts 3 and 4 closed, the motor for the antiwrinkle cycle is energized. But with contacts 5 and 6 open, current is prevented from reaching either the drive motor or timer motor.

The antiwrinkle timer operates on a 5-minute cycle, closing contacts A1 and A2 for about 10 seconds at 5-minute intervals. During the 10-second period both the main drive motor and the regular timer motor are allowed to run. This tumbles the clothing for 10 seconds then stops for 5 minutes.

The main timer will hold contacts 3 and 4 closed for about 5 minutes of its cycle following the end of an automatic dry cycle. However, since the timer motor is energized only 10 seconds every 5 minutes, it will take the main timer motor approximately 2 hours to advance the 5 minutes at the end of the cycle. This means that every 5 minutes (for 2½ hours) after the automatic dry cycle has been completed, the load will be tumbled for 10 seconds.

This additional tumbling will prevent wrinkles from setting in when the operator is not available to remove the clothing at the end of the cycle. When the main timer finally advances 5 minutes, contacts 1 and 2 and contacts 3 and 4 are opened and the cycle is complete.

Electronic Controls

Some dryers use an electronic circuit to control the run time of the machine. This replaces the conventional timer. In these machines, automatic drying time is controlled by the moisture content in the clothes instead of drum temperature.

Resistance of any material is partly determined by its moisture content. Sensors placed in the drum measure the electrical resistance of the clothing. When the moisture content drops to a prescribed level, the electronic control will cause a relay to open the heating circuit and start the cool-down phase of the cycle. Figure 9-9 illustrates the workings of this control.

Sensors placed in the drum make contact with damp clothing, changing the input resistance to the electronic amplifier. When this resistance indicates that further drying is needed, a relay (RY1) is energized and closes the circuit to the heating element and main drive motor. When the drum temperature reaches 120°, the cool-down thermostat closes its contacts. These contacts will remain closed during heat application.

When the resistance of the clothing indicates that they are dry, the electronic amplifier deenergizes the relay and opens

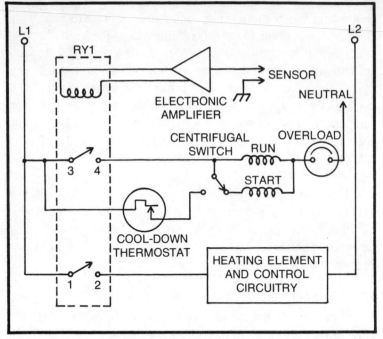

Fig. 9-9. Electronic circuits are sometimes used to control the dryer's operating cycles.

the heating and motor circuits. Notice that the contacts of the cool-down thermostat bypass contacts 3 and 4 of the relay and keep power applied to the motor until the drum cools to 120°. This added tumbling helps prevent wrinkles.

10

Automatic Dryer Installation and Servicing

This chapter discusses the fundamentals of installing and troubleshooting automatic dryers. Installation may or may not be a part of your normal services. Many independent shops contract the installation of machines sold by retail outlets, however. A more important reason for studying installation methods is that proper operation depends on correct installation. To service these appliances you should be able to recognize installation conditions that could affect the operation of the machine.

Exact specifications are contained in the manufacturer's service literature. Always follow these instructions for best results.

INSTALLATION

Many of the considerations that must be observed when installing automatic washers apply to dryer installation as well. Convenience of access, customer's desires, and other personal considerations must all be taken into account. The dryer must be level—front to back and side to side. Most models are equipped with leveling feet for this purpose.

New machines will sometimes require structural changes or modifications to existing electric and gas lines. These changes should be made by licensed personnel. Some service shops require the owners to make these changes before the service technician starts the installation. Other shops sub-contract the plumbing and electrical work, requiring the owner to deal only with the service organization.

Licensing

Most local codes require that all plumbing and electrical modifications be done by persons licensed for that type of work. Your city or county engineer can give you the details on these rules; but, in general, as a service technician your work is limited to the appliance itself. Some shops work out advance agreements with local electrical and plumbing firms to do the necessary structural modifications. This is probably the most satisfactory solution for the small service shop.

All dryer installations must be vented to the outside atmosphere. Some cities insist that vent work requiring wall penetrations be done by licensed servicemen. Familiarity with the codes in your area of operation is the only way to avoid trouble. Know the requirements and limitations.

Electrical Installation

You will be concerned with two entirely different types of electrical installations. The electric dryer will require a special 240V circuit, while the gas dryer requires 120V for motor and control circuit operation. The requirements given here for both installations apply to most automatic dryers. You should consult the manufacturer's instructions for specific details.

Most gas-operated dryers come equipped with a heavy-duty electrical cord and require only a proper service outlet and equipment ground to complete the electrical installation. This service outlet should be capable of supplying 120V AC and should be protected with a 20A circuit breaker or a 15A time-delay fuse. If at all possible a separate service outlet, used by the dryer only, should be provided.

Gas dryers are equipped with a three-prong grounding plug on the power cord. This must be plugged into a properly grounded three-prong receptacle for protection against electrical shock. Where a two-prong receptacle is encountered, it is recommended that a new service receptacle be installed by qualified and licensed electricians. Where changing the receptacle is not possible and where local codes permit, an adapter may be used to connect the three-prong appliance plug to the two-prong receptacle. Always make sure the receptacle is grounded. If not, a separate 18-gage ground wire will have to be run from the wallplate mounting screw to a grounded metal cold water pipe as shown in Fig. 10-1. If necessary, add a jumper around the water meter or any other insulated pipe connections.

Although not required by some codes, it is a good safety practice to run a separate equipment ground from any major appliance to a good ground, such as a cold water pipe. I recommend this added safety on all electrical appliances that are not frequently moved. Although three-wire heavy-duty extension cords are permitted on gas dryer installations by some local codes, the best solution is to avoid the use of any extension cord. After all, the dryer is more or less a permanent installation. Why not make it right to begin with and avoid future problems?

Requirements for an electric dryer are somewhat more complex than those of gas models. A separate 240V AC circuit

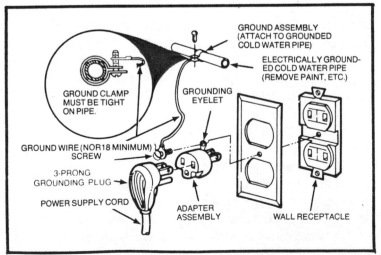

Fig. 10-1. Using the three-prong adapter.

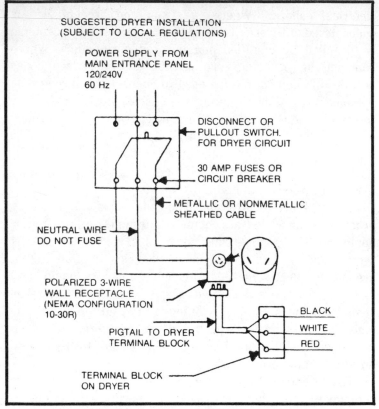

SUGGESTED DRYER INSTALLATION
(SUBJECT TO LOCAL REGULATIONS)

POWER SUPPLY FROM
MAIN ENTRANCE PANEL
120/240V
60 Hz

DISCONNECT OR
PULLOUT SWITCH.
FOR DRYER CIRCUIT

30 AMP FUSES OR
CIRCUIT BREAKER

METALLIC OR NONMETALLIC
SHEATHED CABLE

NEUTRAL WIRE
DO NOT FUSE

POLARIZED 3-WIRE
WALL RECEPTACLE
(NEMA CONFIGURATION
10-30R)

PIGTAIL TO DRYER
TERMINAL BLOCK

TERMINAL BLOCK
ON DRYER

BLACK
WHITE
RED

Fig. 10-2. A polarized three-wire 240V connection.

is required. Both legs of the circuit should be protected with a 30A fuse or circuit breaker. The neutral line should remain unfused. Extension cords should never be used with 240V dryers. In most cases, the power cord for the dryer will not be attached to a new machine. Always follow the manufacturer's instructions when installing these cords. Again, all local codes must be followed on these installations. The procedures explained here are only general guidelines.

Most local codes will permit the separate dryer circuit to terminate at a polarized three-wire wall receptacle and then connect to the dryer with a plug and cord set as shown in Fig. 10-2. This circuit must be equipped with a disconnect or pullout switch and protected with two 30A fuses. Some local codes specify a separate circuit leading directly from the main entrance panel to the dryer. If the dryer circuit is not equipped with a disconnect or pullout switch at the main entrance, one must be installed in the circuit between the main entrance panel and the dryer as shown in Fig. 10-3.

Local codes will also vary on grounding requirements. Some communities do not allow the neutral leg to be grounded in any way. Most electric dryers are shipped with the neutral wire grounded to the frame of the dryer. Where this is not permitted, follow the manufacturer's instructions and the local codes for proper installation. A separate equipment ground is always permitted (and recommended by this service technician). This ground should be a separate wire, not connected to any electrical circuit, running from the main frame of the machine to a cold water pipe. Equipment grounding has been discussed several times in other parts of this book and is not repeated here.

A final word of caution on 240V dryer installations. The wires that serve this appliance should always have *three* conductors, one a bare ground. All three are current carriers

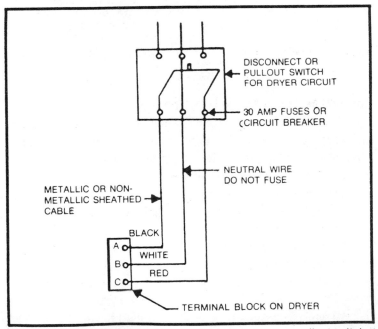

Fig. 10-3. Direct 240V connections require a disconnect or pullout switch.

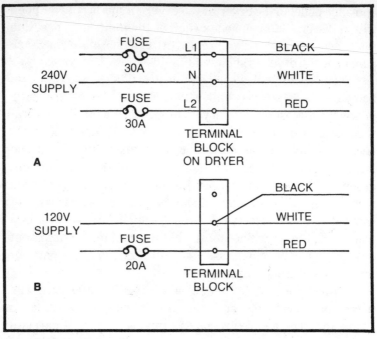

Fig. 10-4. Converting dryers from 240V operation to 120V operation requires the installation of an equipment ground: A shows the connection for 240V operation. B shows the connection for 120V operation.

and should be of equal size and insulation. Again, consult the local codes.

Converting 240V Dryers to 120V Operation

Although this type of installation is not as popular as it once was, you will occasionally encounter a 240V dryer that has been converted for 120V operation. Due to differences in the wiring of the various makes, we will not attempt to give you complete details of such systems. Figure 10-4 shows the wiring of one particular model. The conversion is accomplished by connecting one leg of the internal 240V circuit to the neutral line as shown. A separate equipment ground should be used on all such installations.

Venting

All dryer installations should be vented to expel moisture, lint, and gases from the home. Indoor venting dumps moisture and lint-laden air back into the dwelling. Besides being irritating to the nostrils, lint is very flammable. In addition, byproducts of gas-fired dryers can be very dangerous to breathe and may be explosive. Always vent dryers to the atmosphere. Some codes allow under-the-house venting, but I wouldn't make such installations. I would not want explosive gases or lint to accumulate under my own home. Therefore, I refuse jobs where such installations are required. Also, the dryer should never be exhausted into a chimney.

Figure 10-5 illustrates a good through-the-wall vent installation recommended by this shop. Where necessary, the direction of the venting can be changed by using elbows. The length of the exhaust pipe and the number of elbows should be kept to a minimum to allow the free flow of exhausted materials. When permitted, flexible ducting may be used.

Proper venting is important to obtain maximum drying efficiency. When making a rigid duct connection, the crimped end must be installed downstream of the exhaust flow as shown in Fig. 10-6. If the rigid duct connections are made incorrectly, as shown in Fig. 10-6, lint will collect at each connection and decrease the airflow.

Gas Connections

Gas connections must also be made in accordance with existing codes. Most communities specify the types of pipe,

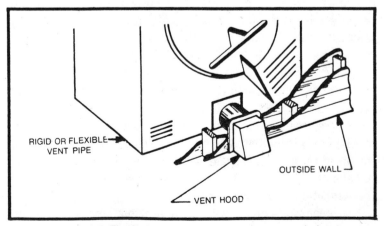

Fig. 10-5. Through-the-wall venting is recommended.

227

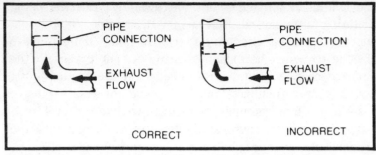

Fig. 10-6. Vent connections should point in the direction of the exhaust flow.

valves, and other fittings that may be used. Never use any of these materials without first consulting local regulations. This is especially true of any flexible tubing.

Your main concern will be with the burner itself. Most burners are designed to operate on any of the three common gases used in the United States—natural, manufactured, and liquified petroleum. Manufacturer's instructions will detail the procedure for changing burner orifices in order to use the different gases.

Be safe. Use only approved materials and procedures. Refer gas pipe work to licensed personnel.

TROUBLESHOOTING

No troubleshooting diagrams, charts, or problem—cause tables will locate every trouble for you. Good troubleshooting practices, however, dictate a logical, step-by-step method of fault isolation. We have found that troubleshooting charts aid in guiding a technician through these logical steps. By following a routine checkout procedure, you insure that no steps are overlooked and reduce the time spent in pinpointing problem areas.

There are three functional areas of operation in any dryer: heating, air circulation, and cycling. Observe the operation of the dryer through each cycle and note the functional operation in each of these areas. If the unit is operating properly in all three functions, there is no problem.

Problems associated with the heat-producing section of an automatic dryer will require one kind of inspection for gas dryers and another kind for electric dryers. If you have trouble producing the required heat with gas burners, refer to the chapter explaining gas dryer operation. Heating problems in electric dryers are best located by using an ohmmeter and the manufacturer's wiring diagram to check the heating element and its control circuitry.

Air-circulation trouble can usually be pinpointed to a plugged vent or mechanical problems associated with the blower, such as a loose belt or pulley. Start at the vent outlet outside the house and see if there is adequate airflow during normal operation. If not, remove the duct connection at the back of the dryer and check there. This will tell you whether the air is stopped inside the dryer or inside the ducting.

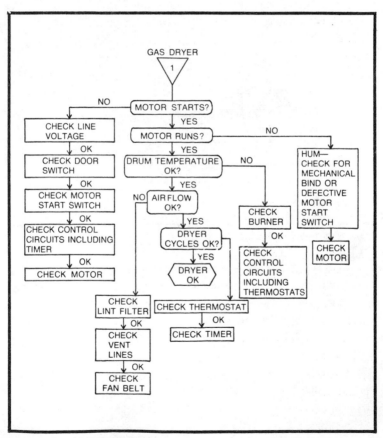

Fig. 10-7. Troubleshooting chart for a gas-powered dryer.

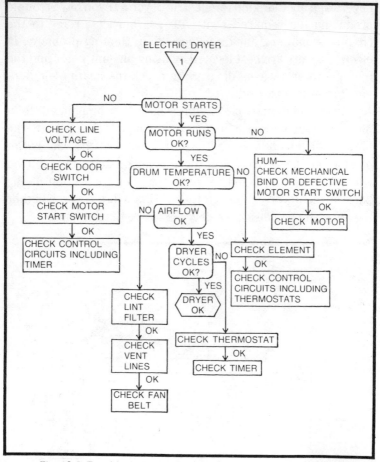

Fig. 10-8. Troubleshooting chart for an electric-powered dryer.

Cycling problems will usually be located in the timer or thermostat circuits. Observe the operation during a typical cycle and try to locate the exact point in the cycle when the failure occurs. This will usually lead you to a particular component. Voltage and resistance readings can confirm your suspicions.

A typical set of troubleshooting tables are given in Figs. 10-7 and 10-8. Always make it a practice to check the obvious first. This will save you troubleshooting time.

Residential
Water
Systems

Clean, clear, potable water in abundance is taken for granted by most homeowners. Little thought is given to the water supply—until something goes wrong. That's where you come in. As an appliance technician you probably will not be called on to do much work on water systems. In fact, many communities require that hot water and water-treatment equipment be installed and serviced by licensed plumbers. Although licensing is not a consideration in our area, my shop does not service this equipment. We feel it is better to leave this work to shops equipped to give complete service —including installation.

Why include this chapter in a book about laundry appliances? Because the water used by laundry appliances has a bearing on their efficiency. Insufficient hot water and rusty or dirty water can result in a poor wash. Chances are, you will get the call before the plumber. You need to know a little about how this equipment works in order to understand how it can affect you. When you get a complaint about poor washing, you'll be the one to determine whether the laundry appliances or water system is at fault.

WATER HEATING AND STORAGE

Although referred to as a heater, the residential water heater not only heats the water as needed but stores it for

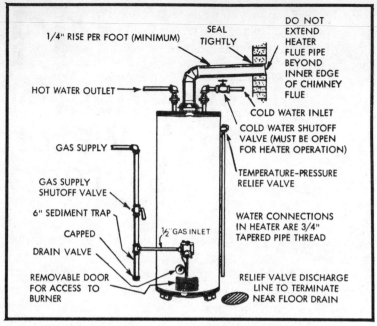

Fig. 11-1. A typical gas-powered water heater installation.

ready use. Size, shape, and capacity will vary from model to model, depending on its intended use. Heating may be either gas or electric. In either case, the water is kept at the selected temperature.

Gas-Powered Water Heaters

A typical gas-powered water heater installation is shown in Fig. 11-1. The location of this appliance should be selected on the basis of accessibility to gas and water lines, ventilation, drainage, and protection against freezing. In addition, the floor supporting the heater should be level and strong enough to support the weight of the heater when it is filled with water. No flammable materials should be stored in the vicinity of gas-powered water heaters.

Proper venting is one of the most important installation considerations. Flue gases must be removed from the dwelling and an adequate supply of fresh air must be available for combustion. The flue run should be as direct as possible with a minimum number of bends and turns. Although the flue may be vented to a chimney, it should not connect to any chimney serving an open fireplace. The heater should not be installed in a small enclosure unless one side of the enclosure is permanently open to an air supply.

Regulations regarding sizes and types of pipe for proper installation are contained in local building codes. They must be followed if you install these units. Most codes require that a combination temperature and pressure-relief valve be installed on the tank. Normally installed in the tank-opening marked *relief valve*, this device is for protection against excessive temperature and pressure in the tank. A drainpipe should lead from the relief valve to a plumbing drain or floor drain.

Figure 11-2 is a cutaway drawing that illustrates the basic operating features of a gas-powered water heater. Cold water

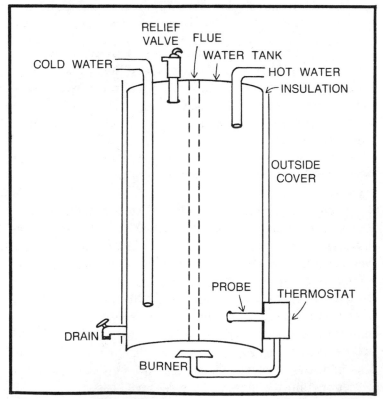

Fig. 11-2. A cutaway drawing of a typical gas-powered water heater.

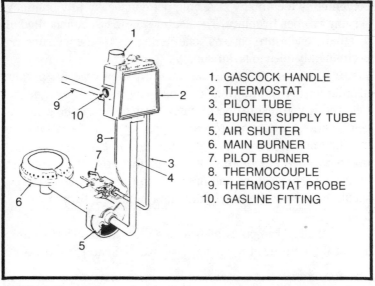

1. GASCOCK HANDLE
2. THERMOSTAT
3. PILOT TUBE
4. BURNER SUPPLY TUBE
5. AIR SHUTTER
6. MAIN BURNER
7. PILOT BURNER
8. THERMOCOUPLE
9. THERMOSTAT PROBE
10. GASLINE FITTING

Fig. 11-3. The gas burner for a water heater is thermostatically controlled.

enters the tank from the water supply system. As hot water is drawn off near the surface of the tank, additional cold water is added at the bottom of the tank. When the water temperature (as sensed by the thermostat probe) drops to a sufficiently low level, the thermostat opens the gas supply to the burner. With the burner ignited, the water temperature will start to rise. When the probe senses that the water temperature has reached the level selected on the thermostat, the burner will cut off.

A more detailed representation of the burner is shown in Fig. 11-3. In many respects, this burner operates like a gas dryer with a standing pilot. The thermostatically controlled valve has a pilot bypass feature that permits the pilot to remain lit at all times. A lever at the top of the thermostat housing provides a means for adjusting the water temperature. When the thermostat probe senses that the water temperature is below the level selected at the thermostat, the valve opens and gas entering the burner is ignited by the pilot. This pilot has a safety feature that turns off the gas to the burner anytime the thermocouple does not sense the presence of a flame at the pilot.

Electric-Powered Water Heaters

Installation requirements for electric-powered water heaters vary somewhat from those for gas-powered water heaters. Venting and air circulation requirements for gas-powered water heaters are very important. Electric models do not use combustion to generate heat, so these requirements need not be considered. General construction features are the same for both heaters.

Specifications for the power connections to electric-powered water heaters will vary somewhat with local codes and the particular unit being installed. In general, however,

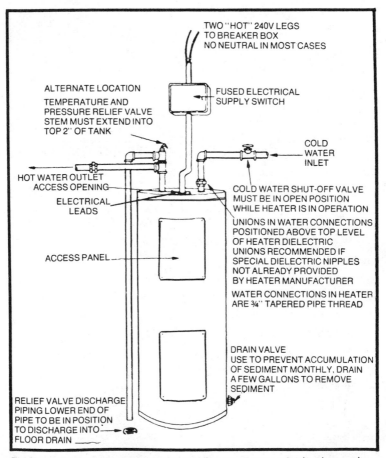

Fig. 11-4. Typical connections for a 240V, three-wire, single-phase, electric-powered water heater.

water heaters are not connected to the neutral lines unless specifically required. A 240V, three-wire, single-phase system connected as shown in Fig. 11-4 is typical. This should be a separate circuit protected by a circuit breaker or fuses in each leg.

Three popular types of heating units are found in electric water heaters. The simplest of these is the *single-element* heater. Operation is very similar to an electric-dryer heating element with a thermostat controlling the current flow through the heater coil as shown in Fig. 11-5A. Figure 11-5B illustrates

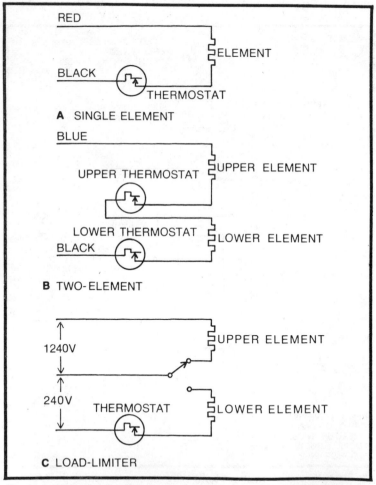

Fig. 11-5. Wiring for electric-powered water heaters: (A) single element; (B) two-element; and (C) the load-limiter.

the operation of a *two-element* (duplex) water heater circuit. Two separate elements, one near the top and the other near the bottom of the tank, are controlled by separate thermostats. Each element operates independently of the other, heating the water in its particular area of the tank. Both will operate at the same time if the entire tank is cold.

The *load-limiter* water heater also has two elements but operates in a somewhat different manner than the duplex circuit. As shown in Fig. 11-5C, the top element operates when contacts 1 and 2 of that thermostat are closed. Once the upper level has been heated, contacts 1 and 2 open, closing contacts 1 and 4. If the water in the lower portion of the tank is cold, the contacts of that thermostat will also be closed causing the lower element to operate. Notice that the lower element cannot operate until the upper level is heated. In this circuit both elements are never on at the same time.

Servicing

Routine servicing of water heaters is limited to occasionally draining the tank to remove sediment. Some manufacturers recommend that a few quarts of water be drawn every month to prevent accumulation of sediment in the bottom of the tank.

Troubleshooting and repairs of water heater burners and heating elements are similar to that for clothes dryers. Voltage and resistance checks will reveal most troubles in an electric water heater, while problems in a gas model can usually be located by observing its operation.

WATER TREATMENT SYSTEMS

In localities where the residential water supply is not sufficiently pure to meet the needs of the consumers, home-water treatment systems are available to correct most deficiencies. Although the water may be potable, it can still contain minerals and other matter harmful to appliances. Deposits from the minerals contained in some water supplies can block water passages and cause corrosion of metalic parts. Water is usually analyzed and tested for four qualities—hardness, iron content, acidity, and turbidity (muddiness).

Water Softeners

Water hardness is measured in mineral grains per gallon. Water softeners are designed to remove enough of these minerals to eliminate the effects of hard water. Too many minerals in the water supply reduce the ability of detergents to release their chemicals. In addition, mineral deposits in mixing valves cause premature failure of these parts.

When you encounter complaints about the lack of sudsing action obtained from detergents, look into the possibility of hard-water problems. I have serviced problems where my customer's reply was, "I have an automatic water softener." A close check of the plumbing sometimes reveals that the softener is installed in the hot water line only. Water softeners must serve both hot and cold water lines to do their job effectively. For example, an automatic washer will use both hot and cold water in the regular cycle of routine washing. Mixing the treated hot water with the untreated cold water will reduce the overall effectiveness of the water softener.

Each softener, as well as the filters discussed here, will have its own peculiarities, depending on the manufacturer. Always consult the manufacturer's service manual on these devices before advising the customer.

Filters

Iron, acidity, and turbidity can all be treated with filter systems. Separate filters can be used for each condition, or a complete water treatment system may be installed.

Iron suspended in the water can cause many washday complaints. In addition to its effects on water taste, iron can stain clothing and plumbing fixtures. Two common types of iron can be found in residential water systems—ferrous iron and ferric iron.

Ferrous iron (sometimes called clear-water iron) can be detected by drawing a glass of water from the faucet and allowing it to stand. At first, the water will appear to be clear. The air starts to oxidize the iron in the water. The water turns yellow and then brown, depending on the amount of ferrous iron. Although a softener can remove ferrous iron in small amounts, a separate filter system must be used to remove large quantities of this mineral.

Ferric iron (sometimes called red-water iron) is clearly visible when the water is drawn from the faucet. The water will appear yellow or reddish-brown. After standing for a period, the suspended particles will settle to the bottom of the glass. A filter system is the only answer to ferric iron problems.

Water is tested for acidity on the pH scale. A measurement less than 6.8 is considered acidic and requires a neutralizing filter to correct the situation. A filter system can also be used to remove sediments such as silt or clay (called turbid water).

Although you may never repair these appliances, you should be aware of their effects on the water that serves appliances you do repair.

Appendix A
All About Motors

Reprinted in part from U.S. Government publication 2257. *Selecting and Using Electric Motors.* by L.H. Soderholm and H.B. Pucket. North Central Region. Agricultural Research Service. Copies of the complete publication may be ordered from the Superintendent of Documents. U.S. Government Printing Office. Washington D.C. 20402. The Federal Stock Number of the publication is 0100-03178.

For proper performance of an electric motor, the supply voltage and frequency at the motor terminals must match the values specified by the manufacturer as closely as possible. Motor performance usually is determined at rated voltage and frequency. Satisfactory performance generally is obtained over a range of plus or minus 10 percent from rated voltage and plus or minus 5 percent from rated frequency.

If you allow the applied voltage or frequency to vary from the nominal values specified on the motor nameplate, there will be changes in the motor torque from the values that are given at rated voltage and frequency. These changes in performance occur because torque developed by the motor is approximately proportional to the square of the voltage and inversely proportional to the square of the frequency.

The successful operation of a motor under running conditions, with the voltage and frequency variations within the allowable range, does not necessarily mean that the motor will start and accelerate the load under these conditions. Limiting values of voltage and frequency at which a motor will start and accelerate a load to running speed depend on the margin between the speed-torque curve of the motor at rated voltage and frequency and the speed-torque curve of the load under starting conditions.

Frequency variation is generally no problem, except for possible operation by standby power units. Low voltage from inadequate wiring or other causes, however, can cause severe problems because motor starting torque may be too low to start and accelerate the load. The section on wiring gives recommended minimum sizes of conductors.

Proper motor voltage is, therefore, particularly important for hard starting loads; at 80 percent

241

of its rated voltage, a motor develops only 64 percent of the torque that is developed at nameplate voltage rating.

Motor Torque Characteristics

An electric motor is simply a device for converting electrical power into mechanical power. Therefore, after the type of power source has been determined, the next step is to determine the motor size required for the load.

To start the load, a certain amount of turning force is required. The motor shaft turning force, or torque, is the turning force available from the motor shaft. A load such as a fan starts easily, and the shaft can be turned by hand (low starting torque). A load such as a filled grain auger starts much harder (high starting torque), and a wrench is needed to rotate the shaft. An electric motor must be selected that will provide the starting torque required by the load as well as the torque necessary to bring the load to operating speed.

When the load is running, a given amount of turning force is required for rotating the load at the desired speed. This power that the motor must put out (horsepower) is proportional to the shaft speed and torque required. Thus, the horsepower required to turn the load determines the size of motor that must be selected. If too small a motor is selected, it will be overloaded and have a short life.

The torque characteristics of a load that a motor is required to start and run are therefore important in the choice of a motor.

Typical torque characteristics of a load and a motor are shown in figure A1. Motor torque must exceed load torque requirements over the entire range of speed if the motor is to properly start and bring the load to operating speed. The motor torque characteristics that are important in matching a motor to a load are defined as follows.

Full-load torque is the turning force that the motor will deliver continuously at rated voltage and speed without exceeding its temperature rating. Full-load torque usually determines the basic rating and therefore size of the motor that must be used.

Starting torque (locked rotor) is the amount of torque that the motor has available at zero speed. Starting torque is important and may dictate the type of motor that must be used.

Breakdown torque is the maximum torque that a motor develops at rated voltage without an abrupt drop in speed. Breakdown torque must be considered in relation to peak intermittent loads that may be encountered.

Pull-up torque is the minimum torque that is developed by the motor during the period of acceleration from zero speed to the speed at which breakdown torque occurs. Pull-up torque is generally of minor importance but must be adequate to accelerate a load up to its operating speed.

Motor Loading

The life of an electric motor is reduced if the motor is overloaded for extended periods. Overload is indicated when the current is

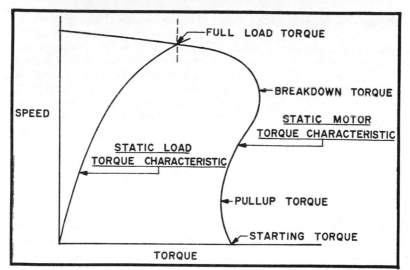

Fig. A1. Typical motor and load torque characteristics.

above the nameplate rating. One method of checking for motor overload is to measure the current drawn by the motor.

Each of the conductors carrying power to the motor is passed through the clamp-on ammeter loop individually to determine the current of the motor under load conditions. Currents should be approximately equal and within the motor nameplate rating for both leads of a single-phase motor and for all three leads of a three-phase motor.

If motor current exceeds the motor nameplate current rating, the motor is most likely overloaded and motor temperature will rise above the rated value. Unless the load is of short duration or the cooling air temperature is below 40° C. (104° F.), motor life will be shortened. The load on the motor should be adjusted so that the current drawn by the motor is within the nameplate rating to obtain normal motor life.

Temperature

Once the required motor torque characteristics are determined, motor temperature ratings should be considered. Both motor insulation and bearings have definite temperature limitations for long successful operation. Generally, however, bearing temperature limits will be met if insulation temperature is kept within the permitted range.

Four insulation systems are available for small induction motors. They are as follows.

System	Maximum hot spot continuous temperature
Class A	105° C. (221° F.)
Class B	130° C. (266° F.)
Class F	155° C. (311° F.)
Class H	180° C. (356° F.)

Temperature limits are established by Underwriters' Laboratories to protect against fire hazards and by the National Electrical Manufacturers Association (NEMA) to assure adequate motor life. Nameplate data gener-

243

ally give the permissible temperature rise above the ambient air or the maximum ambient temperature for motor operation that will keep the hot spot temperature of the motor within the specified value for the insulation system used. Normal maximum ambient temperature is 40° C (104° F.) for most motor ratings.

Farm equipment manufacturers usually recommend the type and size of electric motor needed to operate their equipment. Their recommendations are generally based on the starting, pull-up, breakdown, and running torques required under normal operating conditions and serve as a basis for motor selection. For unusual conditions, consult the equipment manufacturer or power supplier, or both.

Operating Conditions

Electric motors are often operated under adverse conditions where there is dust, dirt, or moisture or where there are explosive mixtures of gas or dust such as in feed or flour mills. Motors are available with different types of enclosures, or housings, for use under specific operating conditions. Selecting the proper type of enclosure is important for the protection of the motor and for safe operation.

Enclosures

Two general types of enclosures are available, open and totally enclosed.

An open motor is one that has ventilating openings that permit the passage of external cooling air over and around the windings of the machine.

A drip-proof enclosure protects a motor from liquids or solids falling zero to 15 degrees downward from vertical. It is designed for indoor use where the air is fairly clean and where there is little danger of splashing liquid.

A splash-proof enclosure protects the motor from liquids or particles that strike the enclosure at angles not greater than 100 degrees downward from vertical. Such motors may be used outdoors but must be protected from the weather.

Totally enclosed motors are those where the enclosure prevents the free exchange of air between the inside and outside of the case but does not make the case completely airtight. They may be cooled by a fan (totally enclosed fan-cooled, TEFC), or by direct radiation and convection of heat through the case (totally enclosed, nonventilated).

Totally enclosed motors are also available in explosion-proof, dust-ignition-proof, and water-proof designs for operation under dirty or wet conditions or where explosive gas or dust mixtures are present.

Bearings

Electric motors are available with either sleeve bearings (fig. A2) or ball bearings. (fig. A 3). Operating conditions determine whether sleeve or ball bearings should be used. Sleeve-bearing motors usually are quieter and cost less than ball-bearing motors but generally require more maintenance.

Sleeve-bearing motors usually are designed to operate only in the horizontal position, a l t h o u g h sleeve bearings are sometimes used in a vertical position in small motors. When sleeve bearings are lubricated with oil, the reservoir must always be toward the bottom of the motor.

Ball-bearing motors may be operated in either a horizontal or vertical position and are better suited for end thrust and control of end play than sleeve-bearing motors. Ball bearings that are normally used in electric motors are designed to absorb some thrust, but if the thrust load is high, special thrust bearings must be used. Also, enclosed motors that are used in wet and dirty conditions usually have ball bearings.

Motor Ratings

Motors of a given horsepower rating are built in a certain size of frame or housing. For standardization, NEMA has assigned the frame size to be used for each integral horsepower motor so that shaft heights and dimensions will be the same to allow motors to be interchanged.

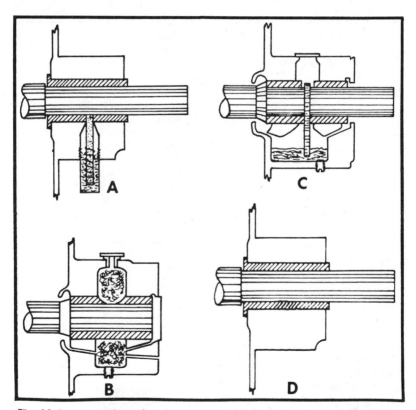

Fig. A2. Low-cost sleeve bearings are suitable for many motor applications. the motor shaft must be mounted horizontally with the oil reservoir underneath. Sleeve bearings will not absorb axial thrust. They may be lubricated by (A) oil wick, (B) yarn, (C) oil ring, or (D) impregnated permanent lubrication.

245

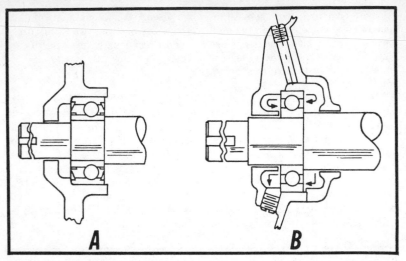

Fig. A3. A motor equipped with ball bearings may be mounted in any position. Ball bearings can take a small amount of axial thrust. Bearings may be either (A) the sealed type, requiring disassembly for relubrication, or (B) the type lubricated with a grease gun.

Motors designed after 1964 are commonly called T-rate motors. These motors are smaller than older motors, which may cause some problems in replacement use.

The smaller size of T-rate motors is the result of closer design tolerances and better magnetic and insulating materials. Also, better insulation allows higher operating temperature within the motor, which makes the old rule no longer valid that you should be able to hold your hand on a motor for 10 seconds or more if the motor is not overheating.

Motor types differ primarily in the amount of starting torque developed and in their starting-current requirements. The type to use depends on the starting requirements of the equipment to be driven and the maximum current that may be drawn from the single-phase power service. Table 1 lists the important characteristics of each type of single-phase motor.

Table 2 shows the motor frame sizes used for various sizes of integral horsepower motors. Shaft diameters for motors with a single straight shaft are shown in table 3. The shaft height of integral horsepower motors may be obtained by dividing the first two numbers of the frame size by 4. Example: The shaft height for the 200-frame-size series is 20 divided by 4, or 5 inches.

The shaft height of fractional horsepower motors may be obtained by dividing the frame size by 16.

Because of tighter design tolerances, the temperature rise of T-rate motors will stay within specifications only if the motor terminal voltage is kept within plus or minus 10 percent of the nameplate rating. The motor winding

Table 1.—Types of single-phase motors and their characteristics

Type	Horsepower ranges	Load-starting ability	Starting current	Characteristics	Electrically reversible	Typical uses
Split-phase	1/20 to ½	Easy starting loads. Develops 150 percent of full-load torque.	High; five to seven times full-load current.	Inexpensive, simple construction. Small for a given motor power. Nearly constant speed with a varying load.	Yes.	Fans, centrifugal pumps; loads that increase as speed increases.
Capacitor-start	1/8 to 10	Hard starting loads. Develops 350 to 400 percent of full-load torque.	Medium, three to six times full-load current.	Simple construction, long service. Good general-purpose motor suitable for most jobs. Nearly constant speed with a varying load.	Yes.	Compressors, grain augers, conveyors, pumps. Specifically designed capacitor motors are suitable for silo unloaders and barn cleaners.
Two-value capacitor	2 to 20	Hard starting loads. Develops 350 to 450 percent of full-load torque.	Medium, three to five times full-load current.	Simple construction, long service, with minimum maintenance. Requires more space to accommodate larger capacitor. Low line current. Nearly constant speed with a varying load.	Yes.	Conveyors, barn cleaners, elevators, silo unloaders.

247

Table 2.—Motor frames for various sizes of 1,800–r.p.m. motors[1]

Kind and size of motor	Motor-frame sizes (NEMA frame series)							
	140	180	200	210	220	250	280	320
	Horsepower							
Single-Phase T-Rate (after 1964)	1 1½	2 3		5 7½				
Single-Phase, U-Rate (1952 to 1964)		1 1½		2 3		5 7½		
Three-Phase, T-Rate (after 1964)	1 1½ 2	3 5		7½ 10		15 20	25 30	40 50
Three-Phase, U-Rate (1952 to 1964)		1 1½ 2		3 5		7½ 10	15 20	25 30
Three-Phase (pre-1952)			1 1½		2 3	5	7½	10 15

[1] The information for this table was taken from NEMA tables MG 1-13.01, 1-13.02, 1-13.01a, and 1-13.02a (1968).

temperature will exceed specifications and motor life will be shortened unless the specified range of motor terminal voltage and load are maintained.

Motors are designed for continuous or limited duty. Those designed for continuous duty will deliver the rated horsepower for an indefinite period of time without overheating. General-purpose motors should always be the continuous-duty type.

Limited-duty motors will deliver rated horsepower for a specified period of time but cannot be operated continuously at the rated load. A typical use of a limited-duty motor is as a silo unloader. A limited-duty motor will operate the unloader satisfactorily for a short period and it costs less than a continuous-duty motor. However, if the operating period is extended, the limited-duty motor will overheat and may burn out prematurely.

Motor nameplates carry the essential information regarding a motor's characteristics. A typical nameplate is shown in figure A4. The information generally given on the nameplate includes the following:

Frame and type.—The NEMA designation for frame designation and type.

Horsepower—The horsepower rating of the motor.

Motor code.—Designated by a letter indicating the starting current required. The higher the locked-rotor kilovolt-ampere (kva), the higher the starting current surge. Table 4 shows the most common letter designations and the locked-rotor kva they represent.

Cycles, or hertz.—The frequency at which the motor is designed to be operated.

Phase.—The number of phases on which the motor operates.

Revolutions per minute (r.p.m.)—The speed of the motor at full load.

Voltage.—The voltage or voltages of operation.

Thermal protection.—An indication of thermal protection provided for the motor, if it is provided.

Amps.—The rated current (amperes) at full load.

Time.—Time rating of the

Table 3.—Shaft diameter for foot-mounted electric motors with a single straight shaft extension[1]

Motor frame size	Shaft diameter inches
143, 145	¾
143T, 145T, 182, 184	⅞
182T, 184T, 213, 215	1⅛
213T, 215T, 254U, 256U	1⅜
254T, 256T, 284TS, 286TS, 284U, 286U, 324S, 326S	1⅝
284T, 286T, 324TS, 326TS, 324U, 326U, 364US, 364TS, 365US, 365TS	1⅞
324T, 326T, 364U, 365U	2⅛
364T, 365T	2⅜

[1] The information for this table was taken from NEMA tables MG1-11.31 and MG1-11.31a.

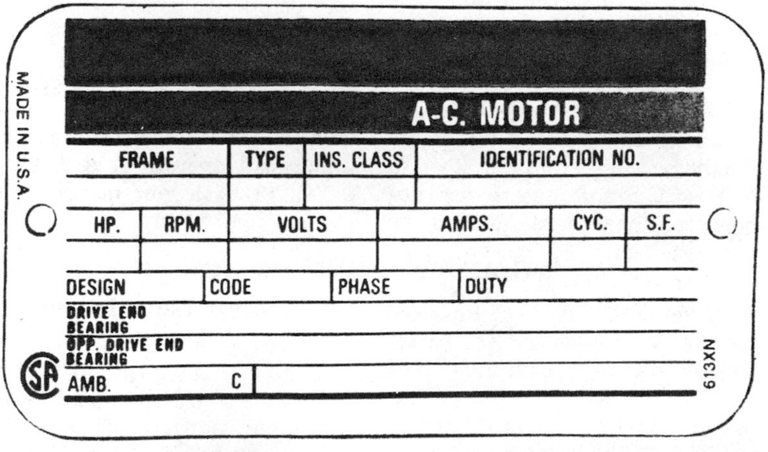

Fig. A4. The motor nameplate gives motor characteristics. The code designation, service factor, time rating, and temperature rise are important considerations in selecting a motor for a given job.

Code letter	Locked rotor[2] kva per horsepower	Horsepower sizes	
		Single-phase	Three-phase
F	5.0 to 5.6		15 up
G	5.6 to 6.3	5	7½ to 10
H	6.3 to 7.1	3	5
J	7.1 to 8.0	1½ to 2	3
K	8.0 to 9.0	¾ to 1	1½ to 2
L	9.0 to 10.0	½	1

[1] The information for this table was taken from NEMA table MG 1-10.37.

[2] Locked rotor kva is equal to the product of line voltage times motor current divided by 1,000 when the rotor is not allowed to rotate; this corresponds to the first power surge required to start the motor. Locked-rotor kva per horsepower range includes the lower figure up to but not including the higher figure.

motor showing the duty rating as continuous or as a specific period of time the motor can be operated.

Ambient temperature, or temperature rise.—The maximum ambient temperature at which the motor should be operated, or the temperature rise of the motor above the ambient air at rated load.

Service factor.—The amount of overload that the motor can tolerate on a continuous basis at rated voltage and frequency.

Insulation class.—A designation of the insulation system used, primarily for convenience in rewinding.

NEMA design.—A letter designation for integral horsepower motors specifying the motor characteristics.

In addition, the bearing designations are often given on the nameplate for both ends of the shaft for convenience in replacement.

Generally a motor with a continuous-duty rating and a 40° C. (72° F.) temperature rise is a good motor capable of operating satisfactorily for an indefinite period of time if properly serviced and operated under normal conditions. However, with the development of improved insulating materials, it is possible to have general-purpose motors which will operate at a rise of 70° C. (158° F.) or more above ambient temperature.

INSTALLATION AND WIRING

Proper installation of an electric motor is essential for satisfactory operation, maximum service, and personal safety. The installation and wiring should conform to the recommendations of the National Electrical Code (NEC) and to any local code that has more restrictive requirements.

Causes of Motor Failure

Motors properly selected and used give many years of satisfactory service. Failures are most often due to overheating, moisture, bearing failure, or starting mechanism failure. Preventive maintenance and proper motor loading are the best insurance against motor failure. Motor life is prolonged by keeping the motor cool, dry, clean, and lubricated.

Overheating.—Heat is one of the most destructive agents causing premature motor failure. Overheating occurs because of motor overloading, low voltage at the motor terminals, excessive ambient temperatures, or poor cooling caused by dirt or lack of ventilation. If heat is not dissipated, insulation failure and possibly bearing failure can ruin a motor.

Moisture.—Moisture should be kept from entering a motor. The proper motor should be chosen for use in a damp environment and it should be covered to protect it from the weather, particularly during periods when it is not used.

Bearing failure.—Bearings should be kept properly lubricated. Bearings may fail in unused motors that are not rotated for extended periods, such as crop driers. Special care in lubrication may be required for these motors.

Starting mechanism failure.—Choice of a well-built motor will help solve this problem. Also, the starting mechanism must be kept free of dirt and moisture, the same as bearings and motor windings.

Mounting

Secure mounting and correct alignment with the load are essential for proper motor performance. The motor should be positioned where it is readily accessible, but not in the way. If possible, the motor should be located so that it will not be exposed to excessive moisture, dust, or abrasive material.

Mount the motor on a smooth, solid foundation and fasten the mounting bolts tightly. If mounted on an uneven base or fastened insecurely, the motor may become misaligned with the load during operation. This will throw unnecessary strain on the frame and bearings, causing rapid wear and overheating. Loose mounting also causes vibration and noise during operation.

Connecting to the Load

Motors may be connected to the load by direct drive, belt and pulley, or chain and sprocket.

Direct drive can be used only when the motor and the driven equipment operate at the same speed. A flexible coupling should be used, and the motor shaft and driven shaft should be in near perfect alignment. This prevents excessive wear of the shaft bearings.

Using a V-belt is the most common and the easiest way of connecting a motor to the load.

High-speed chain drives are used when a positive drive is necessary or when the torque required is more than a V-belt drive can transmit.

Proper belt tension must be maintained. If a belt is too loose, it will slip on the drive pulley, overheat, and wear out quickly. If it is too tight, it will cause the belt and bearings to wear excessively.

To properly tension a V-belt drive, measure the span between shafts as shown in figure 19. Measure the force required to deflect the belt 1/64 inch for each inch of span. The force required should be within the values

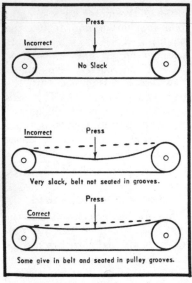

Fig. A5. Adjust V-belt tension correctly.

shown in table 5 for the type of belt used.

Most motors available for farm use operate at about 1,800 r.p.m. Equipment generally operates at much slower speeds. Provision for the required load speed can be made by using the proper size pulley on the driven equipment in relation to the motor pulley. To determine the load-pulley size, multiply the speed of the motor by the diameter of the motor pulley, then divide by the speed of the driven equipment.

Example.—

A load needs to run at 600 r.p.m. The driving motor operates at 1725 r.p.m. and has a 6-inch diameter pulley.

Equipment pulley diameter=
motor speed × motor pulley
diameter
—————————————
equipment speed

Equipment pulley diameter=
$$\frac{1,725 \times 6}{600} = 17.25 \text{ in.}$$

The motor pulley and equipment pulley must be correctly aligned to avoid excessive wear of the belt and bearings

Wiring

For safety, a good ground should be provided to the frame of all electric motors. If an electrical fault develops in the motor or wiring, the ground will prevent hazardous voltages from appearing on the motor frame.

Motors perform best at rated voltage and when adequate wiring is provided to the motor. Operating motors with a terminal supply voltage within the range of rated voltage, and up to plus 10-percent of rated voltage, makes motors less subject to damage during reductions in power system voltage. Adequate voltage also provides better motor performance than that obtained at voltages below the nameplate rating.

Table 6 gives full-load currents of single-phase motors and table 7 gives full-load currents of three-phase motors. Current values shown in these tables should be used for wire size selection unless the motor nameplate current is larger; in that case, use the nameplate current value. Branch-circuit conductors to an individual motor should be selected to carry 125 percent of the full-load current of the motor.

When conductors supply more than one motor on a single circuit, the wire size is determined by taking a current value of 125

Table 5.—Recommended deflection force for V-belt tensioning[1]

V-belt cross section	Small sheave diameter range	Small sheave r.p.m. range	Speed ratio range	Deflection force	
				Minimum	Maximum
type	*inches*			*pounds*	*pounds*
A	3.0 to 3.2		2.0 to 4.0	2.3	3.2
	3.4 to 3.6		2.0 to 4.0	2.5	3.6
	3.8 to 4.2		2.0 to 4.0	2.9	4.2
	4.6 to 7.0		2.0 to 4.0	3.5	5.1
B	4.6		2.0 to 4.0	4.0	5.9
	5.0 to 5.4		2.0 to 4.0	4.5	6.7
	5.6 to 6.4		2.0 to 4.0	5.0	7.4
	6.8 to 9.4		2.0 to 4.0	5.8	8.6
C	7.0		2.0 to 4.0	7.1	10.0
	7.5 to 8.0		2.0 to 4.0	7.9	11.0
	8.5 to 10.0		2.0 to 4.0	9.3	13.0
	10.5 to 16.0		2.0 to 4.0	11.0	16.0
D	12.0 to 13.0		2.0 to 4.0	16.0	24.0
	13.5 to 15.5		2.0 to 4.0	18.0	27.0
	16.0 to 22.0		2.0 to 4.0	21.0	31.0
E	21.6 to 24.0		2.0 to 4.0	33.0	47.0
3V	2.5 to 3.5	1,200 to 3,600	2.0 to 4.0	3.0	4.3
	3.51 to 4.50	900 to 1,800	2.0 to 4.0	3.5	5.3
	4.51 to 6.0	900 to 1,800	2.0 to 4.0	4.3	6.0
5V	7.0 to 9.0	600 to 1,500	2.0 to 4.0	8.8	13.0
	9.1 to 12.0	600 to 1,200	2.0 to 4.0	9.5	14.0
	12.1 to 16.0	400 to 900	2.0 to 4.0	11.0	15.0
8V	12.5 to 17.0	400 to 900	2.0 to 4.0	22.0	31.0
	17.1 to 24.0	200 to 700	2.0 to 4.0	23.0	34.0

[1] Pressure must be applied at midspan perpendicular to the belt. Example: For a span of 32 inches, the measured deflection should be 1/64 × 32, which is 1/2 inch. For a Type A belt with a small sheave diameter of 3 inches, the pressure to produce the 1/2-inch deflection should be 2.3 to 3.2 pounds.

percent of the full-load current of the largest motor plus 100 percent of the current for each additional smaller motor.

The following measures must be provided for in the wiring to motors:

(1) Branch-circuit overcurrent protection to protect the conductors of the motor circuit.

(2) A means to disconnect the motor from the electrical supply.

(3) Motor overcurrent protection to prevent overloading the motor under running conditions.

(4) A controller to stop and start the motor.

Wire Sizes

Tables 8 to 11 show the required wire size for copper and aluminum conductors for single-phase motors and a 2-percent voltage drop. Tables 12 and 13 show equivalent information for three-phase motors. To prevent low voltage from causing improper motor operation, wiring should be selected to limit the voltage drop under full-load conditions to 2 percent for branch circuits and to a total voltage drop of 5 percent for the branch circuit and service wiring combined.

Table 6.—*Full-load currents for single-phase a.c. motors*[1]

Motor horse-power	115 volts		230 volts	
	Full load	125% full load	Full load	125% full load
	amps	amps	amps	amps
1/6	4.4	5.5	2.2	2.8
1/4	5.8	7.2	2.9	3.6
1/3	7.2	9.0	3.6	4.5
1/2	9.8	12.2	4.9	6.1
3/4	13.8	17.2	6.9	8.6
1	16.0	20.0	8.0	10.0
1½	20.0	25.0	10.0	12.5
2	24.0	30.0	12.0	15.0
3	34.0	42.0	17.0	21.0
5	56.0	70.0	28.0	35.0
7½			40.0	50.0
10			50.0	62.0

[1] To obtain full-load currents for 208-volt motors, increase corresponding 230-volt motor full-load current by 10 percent.

Connections

Single-phase, single-speed motors usually have from two to six leads. The number of leads depends on the type of motor and on whether it is a single- or dual-voltage unit.

S p l i t - p h a s e and capacitor motors that are single-voltage and are not reversible (the direction of rotation cannot be changed) have only two leads. Split-phase and capacitor motors that are single-voltage and are reversible have four leads—two for the main winding and two for the auxiliary, or starting, winding.

Dual-voltage capacitor motors have a minimum of six leads— four leads for the main winding and two for the auxiliary winding. For low-voltage operation, all windings are connected in parallel to the line. For high-voltage operation, the main windings are wired in series and the auxiliary winding is connected to

the center leads of the main winding and to one of the supply lines.

The direction of rotation can be changed in split-phase or capacitor motors by reversing the electrical connections of either the main winding or the auxiliary winding to the line (fig. A 6). The terminals may be located on a terminal board or brought out of the motor frame into a terminal box as numbered leads. The wiring diagrams for the specific motor that is being wired must be followed when m a k i n g connections.

Repulsion-start i n d u c t i o n motors and repulsion-induction motors usually are dual-voltage units with four winding leads. For low-voltage operation, the

Table 7.—*Full-load currents for three-phase a.c. motors*[1]

Motor horse-power	Full load	125% full load
	amps	amps
1/2	2.0	2.5
3/4	2.8	3.5
1	3.6	4.5
1½	5.2	6.5
2	6.8	8.5
3	9.6	12.0
5	15.2	19.0
7½	22.0	28.0
10	28.0	35.0
15	42.0	52.0
20	54.0	68.0
25	68.0	85.0
30	80.0	100.0
40	104.0	130.0
50	130.0	162.0
60	154.0	192.0
75	192.0	240.0
100	248.0	310.0
125	312.0	390.0

[1] To obtain full-load currents for 208-volt motors, increase corresponding 230-volt full-load current by 10 percent.

Table 8.—Sizes of copper wire for single-phase, 115-120 volt motors and a 2-percent voltage drop[1]

Load in amps	Minimum allowable wire size			Length of wire to motor in feet													
	Wire in cable, conduit, or earth Types R,T,TW	Types RH, RHW, THW	Bare or covered wire overhead in the air[2]	20	30	40	50	60	80	100	120	160	200	250	300	400	500
				Wire size (AWG or MCM)[3] (Note: Compare the size shown below with the size shown in the column to the left of the double line and use the larger size.)													
5	12	12	10	12	12	12	12	12	12	12	12	10	10	8	8	6	6
6	12	12	10	12	12	12	12	12	12	12	12	10	8	8	8	6	4
7	12	12	10	12	12	12	12	12	12	12	10	10	8	8	6	6	4
9	12	12	10	12	12	12	12	12	12	10	10	8	8	6	6	4	4
10	12	12	10	12	12	12	12	12	10	10	8	8	6	6	4	4	3
12	12	12	10	12	12	12	12	12	10	8	8	6	6	4	4	3	2
14	12	12	10	12	12	12	12	10	10	8	8	6	6	4	4	3	2
16	12	12	10	12	12	12	10	10	8	8	6	6	4	3	3	2	1
18	12	12	10	12	12	12	10	10	8	8	6	6	4	3	3	2	1
20	12	12	10	12	12	10	10	8	8	6	6	4	4	3	2	1	0
25	10	10	10	12	10	10	8	8	6	6	4	4	3	2	1	0	00
30	10	10	10	12	10	8	8	8	6	4	4	3	2	1	1	00	000
35	8	8	10	12	10	8	8	6	6	4	4	3	2	1	1	0	000
40	8	8	10	10	8	8	6	6	4	4	3	2	1	0	0	000	0000
50	6	6	10	10	8	6	6	4	4	3	2	1	0	00	00	0000	250
60	4	6	8	8	8	6	6	4	4	3	2	2	1	0	000	000	300
70	4	4	8	8	6	6	4	4	3	2	1	0	0	00	000	350	350

[1] Use 125 percent of motor nameplate current for single motors.
[2] The wire size in overhead spans must be at least number 10 for spans up to 50 feet and number 8 for longer spans.
[3] AWG is American wire gauge and MCM is thousand circular mil.

255

Table 9.—Sizes of aluminum wire for single-phase, 115-120 volt motors and a 2-percent voltage drop[1]

Load in amps	Minimum allowable wire size			Length of wire to motor in feet													
	Types R,T,TW	Types RH, RHW, THW	Bare or covered wire overhead in the air[2]	20	30	40	50	60	80	100	120	160	200	250	300	400	500
				Wire size (AWG or MCM)[3] (Note: Compare the size shown below with the size shown in the column to the left of the double line and use the larger size.)													
5	12	12	10	12	12	12	12	12	12	10	10	8	8	6	6	4	4
6	12	12	10	12	12	12	12	12	10	10	10	8	6	6	6	4	3
7	12	12	10	12	12	12	12	12	10	10	8	8	6	6	4	4	3
9	12	12	10	12	12	12	12	10	8	8	8	6	6	4	4	3	2
10	12	12	10	12	12	12	10	10	8	8	6	6	4	4	3	2	1
12	12	12	10	12	12	10	10	8	8	6	6	4	4	3	3	1	0
14	12	12	10	12	12	10	10	8	8	6	6	4	4	3	2	1	0
16	10	10	10	12	10	10	8	8	6	6	6	4	4	2	1	0	00
18	10	10	10	12	10	10	8	8	6	6	4	4	3	2	1	0	00
20	10	10	10	12	10	8	8	6	6	4	4	4	3	1	0	00	000
25	10	10	10	10	8	8	6	6	4	4	4	3	2	0	00	000	0000
30	8	8	10	10	8	6	6	6	4	3	3	2	1	00	00	0000	250
35	6	8	10	10	8	6	6	4	4	3	2	1	0	00	00	0000	300
40	6	8	10	8	6	6	4	4	3	2	1	0	0	00	000	250	300
50	4	6	8	8	6	4	4	3	2	1	0	0	00	000	0000	250	300	400
60	2	4	6	6	6	4	3	3	1	0	0	00	000	250	300	350	500
70	2	2	6	6	4	4	3	2	1	0	0	000	000	300	350	500	600

[1] Use 125 percent of motor nameplate current for single motors.
[2] The wire size in overhead spans must be at least number 10 for spans up to 50 feet and number 8 for longer spans.
[3] AWG is American wire gauge and MCM is thousand circular mil.

Table 10.—Sizes of copper wire for single-phase, 230-240 volt motors and a 2-percent voltage drop[1]

Load in amps	Minimum allowable wire size — Wire in cable, conduit, or earth — Types R,T,TW	Minimum allowable wire size — Types RH, RHW, THW	Minimum allowable wire size — Bare or covered wire overhead in the air[2]	20	30	40	50	60	80	100	120	160	200	250	300	400	500
				Length of wire to motor in feet — Wire size (AWG or MCM)[3]. (Note: Compare the size shown below with the size shown in the column to the left of the double line and use the larger size.)													
2	12	12	10	12	12	12	12	12	12	12	12	12	12	12	12	12	12
3	12	12	10	12	12	12	12	12	12	12	12	12	12	12	12	12	10
4	12	12	10	12	12	12	12	12	12	12	12	12	12	12	12	10	10
5	12	12	10	12	12	12	12	12	12	12	12	12	12	12	10	10	8
6	12	12	10	12	12	12	12	12	12	12	12	12	12	10	10	8	8
8	12	12	10	12	12	12	12	12	12	12	12	12	10	8	8	8	6
10	12	12	10	12	12	12	12	12	12	12	12	10	10	8	8	6	6
12	12	12	10	12	12	12	12	12	12	12	12	10	10	8	8	6	6
14	12	12	10	12	12	12	12	12	12	10	10	8	8	6	6	6	4
17	12	12	10	12	12	12	12	12	12	10	10	8	8	6	6	6	4
20	12	12	10	12	12	12	12	12	10	8	8	8	6	6	4	4	3
25	10	10	10	12	12	12	10	10	10	8	8	6	6	4	4	4	2
30	10	10	10	12	12	12	10	10	10	8	6	6	4	4	4	3	2
35	8	8	10	12	12	10	8	8	8	6	6	4	4	3	3	2	1
40	8	8	10	12	12	10	8	8	8	6	6	4	4	3	3	2	1
45	6	6	10	12	10	8	8	8	6	6	6	4	4	3	3	1	0
50	6	6	10	12	10	8	8	8	6	6	4	4	3	2	2	1	0
60	4	6	8	12	10	8	8	8	6	6	4	4	3	2	1	1	0	00
70	4	4	8	10	8	8	6	6	4	4	4	3	2	1	1	0	00	000
80	2	4	6	10	8	6	6	6	4	4	3	3	2	1	0	0	00	000	0000
100	1	3	6	10	8	6	6	4	4	3	2	1	0	00	00	000	0000	250

[1] Use 125 percent of motor nameplate current for single motors.

[2] The wire size in overhead spans must be at least number 10 for spans up to 50 feet and number 8 for longer spans.

[3] AWG is American wire gauge and MCM is thousand circular mil.

Table 11.—Sizes of aluminum wire for single-phase, 230-240 volt motors and a 2-percent voltage drop[1]

Wire size (AWG or MCM)[3]

(Note: Compare the size shown below with the size shown in the column to the left of the double line and use the larger size.)

Load in amps	Minimum allowable wire size			Length of wire to motor in feet													
	Wire in cable, conduit, or earth Types R,T,TW	Types RH, RHW, THW	Bare or covered wire overhead in the air[2]	20	30	40	50	60	80	100	120	160	200	250	300	400	500
2	12	12	10	12	12	12	12	12	12	12	12	12	12	12	12	12	10
3	12	12	10	12	12	12	12	12	12	12	12	12	12	12	12	10	8
4	12	12	10	12	12	12	12	12	12	12	12	12	12	10	10	8	8
5	12	12	10	12	12	12	12	12	12	12	12	12	10	10	8	8	6
6	12	12	10	12	12	12	12	12	12	12	12	10	10	8	8	6	6
8	12	12	10	12	12	12	12	12	12	12	12	10	8	8	6	6	4
10	12	12	10	12	12	12	12	12	12	12	10	8	8	6	6	4	4
12	12	12	10	12	12	12	12	12	10	10	10	8	6	6	4	4	3
14	10	12	10	12	12	12	12	10	10	8	8	8	6	6	4	4	3
17	10	10	10	12	12	12	12	10	8	8	8	6	6	4	4	3	2
20	10	10	10	12	12	12	12	10	8	6	6	6	4	4	3	2	1
25	10	10	10	12	12	12	10	8	6	6	6	4	4	3	2	1	0
30	8	8	10	12	12	10	10	8	6	6	6	4	3	2	2	0	00
35	6	8	10	12	10	10	8	6	6	6	4	4	3	2	1	0	00
40	6	8	10	12	10	8	8	6	6	4	4	3	2	1	0	00	000
45	4	6	10	10	8	8	8	6	4	4	4	3	2	1	0	00	000
50	4	6	8	10	8	8	6	6	4	4	3	2	1	0	00	000	0000
60	2	4	6	10	8	6	6	4	4	3	3	1	1	00	00	0000	250
70	2	4	6	8	6	6	6	4	3	3	2	1	0	00	0000	0000	300
80	1	2	6	8	6	6	4	3	2	2	1	0	00	000	0000	250	300
100	0	1	4	8	6	4	4	3	2	1	0	00	000	0000	250	300	400

[1] Use 125 percent of motor nameplate current for single motors.

[2] The wire size in overhead spans must be at least number 10 for spans up to 50 feet and number 8 for longer spans.

[3] AWG is American wire gauge and MCM is thousand circular mil.

Table 12.—Sizes of copper wire for three-phase, 230-240 volt motors and a 2-percent voltage drop[1]

Load in amps	Minimum allowable wire size			Length of wire to motor in feet													
	Wire in cable, conduit, or earth — Types R,T,TW	Types RH, RHW, THW	Bare or covered wire overhead in the air[2]	20	30	40	50	60	80	100	120	160	200	250	300	400	500
				Wire size (AWG or MCM)[3] (Note: Compare the size shown below with the size shown in the column to the left of the double line and use the larger size.)													
2	12	12	10	12	12	12	12	12	12	12	12	12	12	12	12	12	12
3	12	12	10	12	12	12	12	12	12	12	12	12	12	12	12	12	12
4	12	12	10	12	12	12	12	12	12	12	12	12	12	12	12	12	10
5	12	12	10	12	12	12	12	12	12	12	12	12	12	12	12	10	10
6	12	12	10	12	12	12	12	12	12	12	12	12	12	12	10	10	8
8	12	12	10	12	12	12	12	12	12	12	12	12	12	10	10	8	8
10	12	12	10	12	12	12	12	12	12	12	12	12	10	10	8	8	6
12	12	12	10	12	12	12	12	12	12	12	12	12	10	8	8	6	6
15	12	12	10	12	12	12	12	12	12	12	12	10	8	8	6	6	4
20	12	12	10	12	12	12	12	12	12	12	10	10	8	6	6	4	4
25	10	10	10	12	12	12	12	12	10	10	8	8	6	6	4	4	3
30	10	10	10	12	12	12	10	10	10	8	8	6	6	4	4	3	2
35	8	8	10	12	12	12	10	10	8	8	8	6	4	4	4	2	1
40	8	8	10	12	12	12	10	10	8	8	6	6	4	4	3	2	1

259

Table 12.—Sizes of copper wire for three-phase, 230-240 volt motors and a 2-percent voltage drop—continued

Load in amps	Minimum allowable wire size			Length of wire to motor in feet													
	Types R,T,TW	Types RH, RHW, THW — Wire in cable, conduit, or earth	Bare or covered wire overhead in the air²	20	30	40	50	60	80	100	120	160	200	250	300	400	500
				Wire size (AWG or MCM)³ (Note: Compare the size shown below with the size shown in the column to the left of the double line and use the larger size.)													
45	6	8	10	12	12	10	10	8	8	6	6	4	4	3	2	1	0
50	6	6	10	12	12	10	10	8	8	6	6	4	4	3	2	1	0
60	4	6	8	12	10	10	8	8	6	6	4	4	3	2	1	0	00
70	4	4	8	12	10	8	8	8	6	4	4	3	2	1	1	00	000
80	**3**	4	6	12	10	8	8	6	6	4	4	3	2	1	0	00	000
100	1	3	6	10	8	8	6	6	4	4	3	2	1	0	00	000	000
120	0	1	4	10	8	6	6	4	4	3	2	1	0	00	000	000	0000
150	000	0	3	8	6	4	4	3	2	1	0	0	0	00	000	0000	250
180	0000	000	1	8	6	4	4	3	2	1	0	00	00	000	0000	250	300
210	250	0000	0	8	6	4	4	3	2	1	0	00	000	0000	250	350	400
240	300	250	00	6	4	4	3	2	1	0	00	000	0000	250	300	400	500

¹ Use 125 percent of motor nameplate current for single motors.

² The wire size in overhead spans must be at least number 10 for spans up to 50 feet and number 8 for longer spans.

³ AWG is American wire gauge and MCM is thousand circular mil.

Table 13.—Sizes of aluminum wire for three-phase, 230-240 volt motors and a 2-percent voltage drop[1]

(Note: Compare the size shown below with the size shown in the column to the left of the double line and use the larger size.)

Wire size (AWG or MCM)[3]

Load in amps	Minimum allowable wire size			Length of wire to motor in feet													
	Wire in cable, conduit, or earth		Bare or covered wire overhead in the air[2]	20	30	40	50	60	80	100	120	160	200	250	300	400	500
	Types R,T,TW	Types RH, RHW, THW															
2	12	12	10	12	12	12	12	12	12	12	12	12	12	12	12	12	12
3	12	12	10	12	12	12	12	12	12	12	12	12	12	12	12	10	10
4	12	12	10	12	12	12	12	12	12	12	12	12	12	12	10	10	8
5	12	12	10	12	12	12	12	12	12	12	12	12	12	10	10	8	8
6	12	12	10	12	12	12	12	12	12	12	12	12	10	10	8	8	6
8	12	12	10	12	12	12	12	12	12	12	12	10	10	8	8	6	6
10	12	12	10	12	12	12	12	12	12	12	10	10	8	8	6	6	6
12	12	12	10	12	12	12	12	12	12	10	10	8	8	6	6	4	4
15	12	12	10	12	12	12	12	12	10	10	8	8	6	6	4	4	4
20	10	10	10	12	12	12	12	10	10	8	8	6	6	4	4	3	3
25	10	10	10	12	12	12	10	10	8	8	6	6	4	4	3	2	2
30	8	8	10	12	12	10	8	8	8	6	6	4	4	3	2	1	1
35	6	8	10	12	12	10	8	8	6	6	6	4	3	2	2	0	0
40	6	8	10	12	10	8	8	8	6	6	4	4	3	2	1	0	00
45	4	6	10	12	10	8	8	6	6	4	4	3	2	1	0	00	000

Table 13.—Sizes of aluminum wire for three-phase, 230-240 volt motors and a 2-percent voltage drop[1] continued

Load in amps	Minimum allowable wire size			Length of wire to motor in feet													
	Types R,T,TW	Types RH, RHW, THW	Bare or covered wire overhead in the air[2]	20	30	40	50	60	80	100	120	160	200	250	300	400	500
				Wire size (AWG or MCM)[3] (Note: Compare the size shown below with the size shown in the column to the left of the double line and use the larger size.)													
50	4	6	8	12	10	8	8	6	6	4	4	3	2	1	0	00	000
60	3	4	6	10	8	8	6	6	4	4	3	2	1	0	00	000	0000
70	2	3	6	10	8	6	6	6	4	3	3	1	0	00	00	0000	250
80	1	2	6	10	8	6	6	4	4	3	2	1	0	00	000	0000	300
100	0	1	4	8	6	6	4	4	3	2	1	0	00	000	0000	300	350
120	000	00	2	8	6	4	4	3	2	1	0	00	000	0000	250	350	400
150	0000	000	1	6	4	4	3	2	1	0	00	00	000	250	300	400	500
180	300	0000	0	6	4	3	2	1	0	00	00	000	0000	300	350	500	600
210	350	300	00	6	4	3	2	1	0	00	000	0000	300	350	400	600	700
240	500	350	000	4	3	2	1	0	00	000	0000	250	350	400	500	700	800

[1] Use 125 percent of motor nameplate current for single motors.
[2] The wire size in overhead spans must be at least number 10 for spans up to 50 feet and number 8 for longer spans.
[3] AWG is American wire gauge and MCM is thousand circular mil.

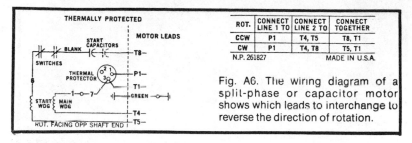

ROT.	CONNECT LINE 1 TO	CONNECT LINE 2 TO	CONNECT TOGETHER
CCW	P1	T4, T5	T8, T1
CW	P1	T4, T8	T5, T1

N.P. 261827   MADE IN U.S.A.

Fig. A6. The wiring diagram of a split-phase or capacitor motor shows which leads to interchange to reverse the direction of rotation.

main windings are wired in parallel. For high-voltage operation, the windings are wired in series. No change in the rotor-brush connections is necessary for operation at either voltage.

MOTOR PROTECTION AND CONTROL

Protection

Motors m u s t be protected against both excessive current and excessive winding temperature caused by faults, overloads, or low supply voltage.

An overloaded motor draws excessive current from the line. This causes overheating that destroys the winding insulation and causes bearings to fail. Maximum temperature at which a motor can operate depends on its construction and the type of insulation used for the windings.

Motor current required for starting usually will be two to eight times that for running at full load. Short-circuit protection devices must be able to carry this high current for a short time.

Branch circuit fuses or circuit breakers do not adequately protect the motor; they primarily protect the circuit wires and give limited protection to the motor for short-circuit conditions only. Additional overcurrent protection for the motor is therefore needed.

Controls

Controls for electric motors vary from a simple on-off toggle switch to complex automatic systems. Motor controls have two purposes: (1) start and stop the motor and (2) protect the motor from damage caused by excessive current. Only simple controls are discussed in this bulletin. Advice on more complex systems should be obtained from your local electrician or power supplier.

Manually Operated Switches

Manual switches are used most often to control small motors of one-half horsepower or less. These switches are low-cost devices. They are available with a built-in overcurrent cut-out that can be sized to the current demand of a particular motor. If the switch does not provide overcurrent protection, these motors must be provided with a suitable overload device such as a dual element fuse, sized specifically to protect the motor, or by a thermal overload built into the motor. Typical manual switch motor-control circuits are shown in figure A7.

Control switches for electric motors must be able to withstand the high starting current and the arcing that occurs when the circuit is opened.

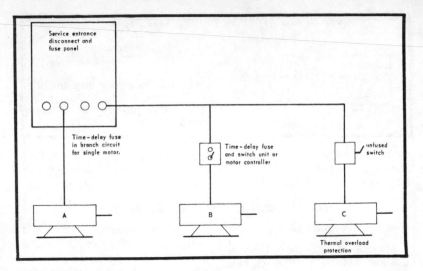

Fig. A7. Typical manual-switch, motor-control circuits.

T-rated, tumbler-type l i g h t switches should not be used to control electric motors. They can withstand the high starting current but are not equipped with arc quenchers and usually burn out quickly.

Magnetic Motor Starters

A magnetic motor starter is the best kind to use for controlling a motor. This type starter should be used for all motors larger than one horsepower and is essential in automatic control systems. A magnetic coil allows operation from either local or remote locations and will remove the motor from the line if there is a loss of power. Built-in thermal or other type overload elements provide overcurrent protection for the motor and should be sized for the specific motor controlled.

There are many ways to connect the control circuits of magnetic motor starters. A commonly used circuit for a 230-volt single-phase motor is shown in figure A 8. A more complex circuit is shown in figure A 9, which provides a sequenced start for three motors. Simultaneous stopping of all motors is provided and overload contacts are interlocked to provide shutdown of all motors if any one of the motors is shut down because of overload.

Generally, motors should not be allowed to restart automatically after a loss of power. If automatic operation is necessary, provision should be made for random restarting to prevent the excessive voltage drop in the wiring that would occur if all motors came on at one time. This can be accomplished by including a low-cost time-delay relay in the magnetic motor starter-control starter as shown in figure A10. This random restart feature is especially desirable for large-horsepower motors, such as those often found in crop-drying fans.

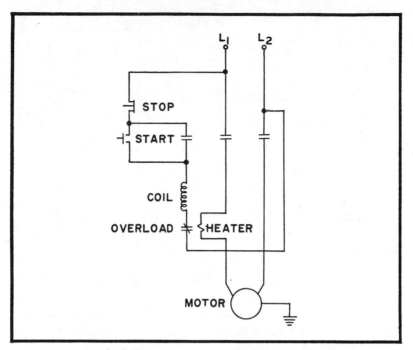

Fig. A8. Single-phase motor starter circuit with 240-volt holding cell.

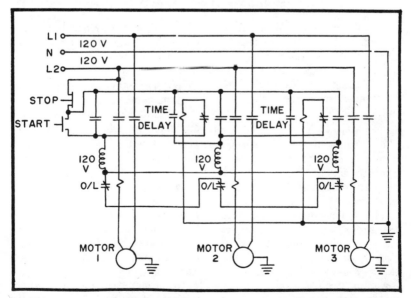

Fig. A9. Sequenced motor starting circuit with instantaneous stopping of all motors and interlocked overload relays.

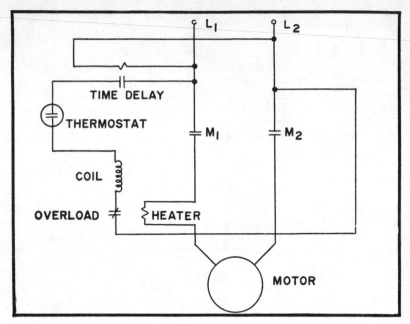

Fig. A10. Circuit for random restarting of motors under automatic control after a loss of power.

SERVICING AND REPAIRS

A well-made and properly installed electric motor requires less maintenance than many other types of equipment. However, for the best and most economical performance, periodic servicing is required.

The service operations listed should be performed at least once a year or more often if the motor operates under severe heat, cold, or dust conditions.

(1) Remove dust and dirt from the air passages and cooling surfaces of the motor to insure proper cooling. Plugged air passages of an open motor, or a coating of dust on a totally enclosed motor, will cause the motor to overheat under normal operation.

(2) Check bearings for wear. Excessive side or end play may cause the motor to draw higher than normal starting current, develop less starting torque, and may damage the motor.

(3) Make sure the motor shaft turns freely. Tight or misaligned bearings will cause the motor to overheat.

(4) Lubricate the motor according to the manufacturer's specifications. Do not overlubricate. Too much lubricant is as bad as too little.

(5) Check all wiring for frayed or bare spots. Repair or replace as needed.

(6) Clean the starting-switch contacts of split-phase and capacitor motors and the commutator

and brushes in wound-rotor (repulsion-type) motors. Use very fine sand paper; *do not use emery cloth.*

(7) Replace worn brushes and make sure the brush-lifting and shorting-ring action works smoothly in wound-rotor motors.

(8) Check belt pulleys to be sure they are secure on their shafts. Align the belts and pulleys carefully. Improper alignment causes excessive wear on belts and pulleys. Check and adjust belt tension. Replace belts that are badly worn.

Properly installed and maintained electric motors should give trouble-free service for many years. Occasionally, however, a motor may give trouble or fail to operate. Some repairs require the services of an experienced electrician or motor serviceman; others can be made by the operator.

Table 14 lists some common motor troubles, their causes, and the methods of repair. Table 15 gives additional information for troubles peculiar to wound-rotor motors.

Caution: Do not attempt to service or repair an electric motor until it has been disconnected from the circuit.

Appendix B
Electrical Symbols

NAME	SYMBOL
BATTERY	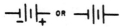
CAPACITOR, FIXED	
CIRCUIT BREAKERS AIR CIRCUIT BREAKER	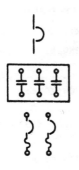
THREE-POLE POWER CIRCUIT BREAKER (SINGLE THROW)(WITH TERMINALS)	
THERMAL TRIP AIR CIRCUIT BREAKER	
COILS NON-MAGNETIC CORE-FIXED	
MAGNETIC CORE-FIXED	

NAME	SYMBOL
MAGNETIC CORE—ADJUSTABLE TAP OR SLIDE WIRE	
OPERATING COIL	
BLOWOUT COIL	
BLOWOUT COIL WITH TERMINALS	
SERIES FIELD	
SHUNT FIELD	
COMMUTATING FIELD	

CONNECTIONS (MECHANICAL)

MECHANICAL CONNECTION OF SHIELD	
MECHANICAL INTERLOCK	
DIRECT CONNECTED UNITS	

CONNECTIONS (WIRING)

ELECTRIC CONDUCTOR—CONTROL	
ELECTRIC CONDUCTOR—POWER	
JUNCTION OF CONDUCTORS	
WIRING TERMINAL	
GROUND	
CROSSING OF CONDUCTORS — NOT CONNECTED	
CROSSING OF CONNECTED CONDUCTORS	
JOINING OF CONDUCTORS — NOT CROSSING	

CONTACTS (ELECTRICAL)

NORMALLY CLOSED CONTACT (NC)	

NAME	SYMBOL

NORMALLY OPEN CONTACT (NO)

_NO_ CONTACT WITH TIME CLOSING (TC) FEATURE
 TC

_NC_ CONTACT WITH TIME OPENING (TO) FEATURE
TO

NOTE: NO (NORMALLY OPEN) AND NC (NORMALLY CLOSED) DESIGNATES THE POSITION OF THE CONTACTS WHEN THE MAIN DEVICE IS IN ② THE DE-ENERGIZED OR NONOPER-ATED POSITION.

CONTACTOR, SINGLE-POLE, ELECTRICALLY OPERATED, WITH BLOWOUT COIL

NOTE: FUNDAMENTAL SYMBOLS FOR CONTACTS, COILS, MECHANICAL CONNECTIONS, etc., ARE THE BASIS OF CONTACTOR SYMBOLS

FUSE

INDICATING LIGHTS

INDICATING LAMP WITH LEADS

INDICATING LAMP WITH TERMINALS

INSTRUMENTS

AMMETER, WITH TERMINALS   (A)   OR

VOLTMETER, WITH TERMINALS   (V)   OR

WATTMETER, WITH TERMINALS   OR

MACHINES (ROTATING)

MACHINE OR ROTATING ARMATURE

SQUIRREL-CAGE INDUCTION MOTOR

NAME	SYMBOL

WOUND-ROTOR INDUCTION MOTOR
OR GENERATOR

SYNCHRONOUS MOTOR, GENERATOR
OR CONDENSER

D-C COMPOUND MOTOR OR GENERATOR

NOTE: COMMUTATING, SERIES, AND SHUNT
FIELDS MAY BE INDICATED BY
1, 2 AND 3 ZIGZAGS RESPECTIVELY.
SERIES AND SHUNT COILS MAY BE
INDICATED BY HEAVY AND LIGHT
③ LINES OR 1 AND 2 ZIGZAGS RE-
SPECTIVELY.

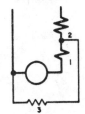

WINDING SYMBOLS

THREE PHASE WYE (UNGROUNDED)

THREE PHASE WYE (GROUNDED)

THREE PHASE DELTA

NOTE: WINDING SYMBOLS MAY BE SHOWN
IN CIRCLES FOR ALL MOTOR AND
GENERATOR SYMBOLS.

RECTIFIER, DRY OR ELECTROLYTIC,
FULL WAVE

FULL WAVE

RELAYS

OVERCURRENT OR OVERVOLTAGE RELAY
WITH 1 *NO* CONTACT

OR

THERMAL OVERLOAD RELAY HAVING
2 SERIES HEATING ELEMENTS AND
1 *NC* CONTACT

OR

RESISTORS

RESISTOR, FIXED, WITH LEADS

RESISTOR, FIXED, WITH TERMINALS

NAME	SYMBOL

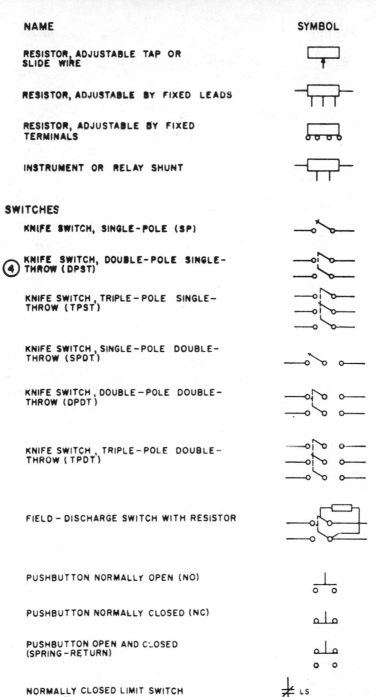

RESISTOR, ADJUSTABLE TAP OR SLIDE WIRE

RESISTOR, ADJUSTABLE BY FIXED LEADS

RESISTOR, ADJUSTABLE BY FIXED TERMINALS

INSTRUMENT OR RELAY SHUNT

SWITCHES

KNIFE SWITCH, SINGLE-POLE (SP)

④ **KNIFE SWITCH, DOUBLE-POLE SINGLE-THROW (DPST)**

KNIFE SWITCH, TRIPLE-POLE SINGLE-THROW (TPST)

KNIFE SWITCH, SINGLE-POLE DOUBLE-THROW (SPDT)

KNIFE SWITCH, DOUBLE-POLE DOUBLE-THROW (DPDT)

KNIFE SWITCH, TRIPLE-POLE DOUBLE-THROW (TPDT)

FIELD-DISCHARGE SWITCH WITH RESISTOR

PUSHBUTTON NORMALLY OPEN (NO)

PUSHBUTTON NORMALLY CLOSED (NC)

PUSHBUTTON OPEN AND CLOSED (SPRING-RETURN)

NORMALLY CLOSED LIMIT SWITCH CONTACT

273

NAME	SYMBOL

NORMALLY OPEN LIMIT SWITCH CONTACT $\quad$ LS

THERMAL ELEMENT

TRANSFORMERS

I PHASE TWO-WINDING TRANSFORMER

⑤ AUTOTRANSFORMER SINGLE-PHASE

Index